Lecture Notes in Mathematics

Edited by A. Dold and B. Eckmann

1368

T0222871

Reinhold Hübl

Traces of Differential Forms and Hochschild Homology

Springer-Verlag

Berlin Heidelberg New York London Paris Tokyo

Author

Reinhold Hübl
Purdue University, Department of Mathematics
Mathematical Sciences Building
West Lafayette, IN 47907, USA

Mathematics Subject Classification (1980): 13 D 99, 14 F 10, 16 A 61, 32 A 27

ISBN 3-540-50985-2 Springer-Verlag Berlin Heidelberg New York
ISBN 0-387-50985-2 Springer-Verlag New York Berlin Heidelberg

© Springer-Verlag Berlin Heidelberg 1989
Printed in Germany

Printing and binding: Druckhaus Beltz, Hemsbach/Bergstr.
2146/3140-543210

Contents

Introduction.

Differential forms and their traces have a long tradition in mathematics. In analysis they were used to study finite coverings of Riemann surfaces. Algebraic analogues appeared first in the theory of algebraic function fields in one variable, for example in M. Deuring's "Theorie der Korrespondenzen algebraischer Funktionenkörper II" (Journal für die reine und angewandte Mathematik, Bd. 183, 1941) and in C. Chevalley's "Introduction to algebraic functions in one variable." These papers are strongly oriented towards the corresponding analytic theory. In particular they only consider differentials on function fields with respect to the field of coefficients, so that strong results could be achieved only in case of separable field extensions.

Later E. Kunz and H. J. Nastold studied one–dimensional function fields K/k, emphasizing the inseparable case. They showed that it is necessary to replace the field of coefficients k of K by a suitable subfield k_0 of k, a so–called admissible field for K/k, and to look at differential forms of K/k_0 in order to get for instance a general duality theorem for function fields of dimension 1 ([K_3], [Na]). However they still only considered traces of differential forms of top degree.

In 1964 E. Kunz gave a general treatment of differential algebras of field extensions. Given a finite field extension L/K and a differential algebra Ω of K he showed that the canonical trace $L \to K$ can be extended in a natural way to a morphism $\sigma^\Omega_{L/K} : \Omega_L \to \Omega$, where Ω_L is the universal L–extension of Ω ([KD], §3). Furthermore he proved that this map is non–trivial, even if L/K is inseparable, and he gave an axiomatic description of this map ([K_4], §2).

Using the global duality machinery developed in [RD] A. Grothendieck derived a general theory of trace for finite morphisms of noetherian schemes, and he showed:

If S/R is a finite algebra and R/k and S/k are smooth algebras of relative dimension n then there exists a natural map "trace": $\Omega^n_{S/k} \to \Omega^n_{R/k}$ ([RD]; III. (8.5)).

In order to find constructive approaches to these traces in the spirit of Kunz's work for field extensions, it turned out to be useful to formulate the problem as follows (c.f. [KD], §16):

For which classes X of pairs $(S/R, \Omega)$, consisting of a finite locally free algebra S/R and a differential algebra Ω of R do there exist trace maps

$$\sigma^{\Omega}_{S/R} : \Omega_S \to \Omega \qquad ((S/R, \Omega) \text{ in } X)$$

where Ω_S is the universal S–extension of Ω ([**KD**], §3), such that the following conditions ("trace axioms") are satisfied:

TR1 (Linearity)

If we consider Ω_S as a left Ω–module via the canonical map $\Omega \to \Omega_S$, then $\sigma^{\Omega}_{S/R}$ is Ω–linear and homogeneous of degree 0.

TR2 (Relation to the canonical trace)

The restriction $\sigma^{\Omega}_{S/R} | \Omega^0_S : S \to R$ to the elements of degree 0 is the canonical trace $\sigma_{S/R}$ ([**KD**], app. F).

TR 3 (base change)

If $\alpha : R \to R'$ is a ring homomorphism, if $S' := R' \otimes_R S$, if $\Omega \to \Omega'$ is an α–homomorphism of Ω into a differential algebra Ω' of R', and if $(S'/R', \Omega')$ is in X, too, then the following diagram commutes:

$$
\begin{array}{ccc}
\Omega_S & \longrightarrow & \Omega'_{S'} \\
\downarrow{\scriptstyle \sigma^{\Omega}_{S/R}} & {\scriptstyle \sigma^{\Omega'}_{S'/R'}} \downarrow & \\
\Omega & \longrightarrow & \Omega'
\end{array}
$$

TR 4 (Direct products)

If $S = S_1 \times \cdots \times S_h$ is a direct product of algebras S_i/R and if $(S_i/R, \Omega)$ is in $X (i = 1, \cdots, h)$, then for each $\omega = (\omega_1, \cdots, \omega_h) \in \Omega_{S_1} \times \cdots \times \Omega_{S_h} = \Omega_S$ we have

$$\sigma^{\Omega}_{S/R}(\omega) = \sum_{i=1}^{h} \sigma^{\Omega}_{S_i/R}(\omega_i).$$

TR 5 (Transitivity)

If $(S/R, \Omega)$, $(T/S, \Omega_S)$ and $(T/R, \Omega)$ are in X, then

$$\sigma^{\Omega}_{T/R} = \sigma^{\Omega}_{S/R} \circ \sigma^{\Omega_S}_{T/S}.$$

TR 6 (Differentiation)

Let d be the differentiation of Ω and let d' be the differentiation of Ω_S. Then

$$d \circ \sigma^{\Omega}_{S/R} = \sigma^{\Omega}_{S/R} \circ d'.$$

TR 7 (Logarithmic derivative)

If $a \in S$ is a unit, then

$$\sigma_{S/R}\left(\frac{da}{a}\right) = \frac{dn_{S/R}(a)}{n_{S/R}(a)},$$

where $n_{S/R} : S \to R$ is the canonical norm ([**KD**], app. F).

In addition to TR 1–TR 7 in this paper the following axiom will be added:

TR 8 (Local algebras)

Let S/R be a local algebra, m_S resp. m_R the maximal ideal of S resp. R and $K := R/m_R, L := S/m_S$. Furthermore set $n := rg_R(S) \cdot \dim_K(L)^{-1}$. If $(S/R, \Omega)$ and $(L/K, \Omega_K)$ are in X, then the following diagram commutes:

$$
\begin{array}{ccc}
\Omega_S & \longrightarrow & \Omega_L \\
{\scriptstyle \sigma^\Omega_{S/R}} \downarrow & n \cdot \sigma^{\Omega_K}_{L/K} \downarrow & \\
\Omega & \longrightarrow & \Omega_K.
\end{array}
$$

In particular it holds: $\sigma^\Omega_{S/R}(m_S, dm_S) \subseteq (m_R, dm_R)$.

Traces of differential forms have been constructed for the following classes, although the axioms TR 1–TR 8 haven't been proved in some cases:

Class I contains all pairs $(S/R, \Omega)$, where S/R is finite and projective, Ω is a differential algebra of R and $\mathbf{Q} \subseteq R$. (B. Angéniol, [**A**], (7.1.2.) and [**AL**], §6.)

Class II contains all pairs $(S/R, \Omega)$, where R is a finite direct product of fields, S/R is a finite algebra, and Ω is a differential algebra of R. (E. Kunz, [**KD**], §16, ex. 5.)

Class III contains all pairs $(S/R, \Omega)$, where S/R is finite and projective, 2 is not a zero–divisor on R or R is reduced and noetherian, and Ω is an exterior differential algebra of R, such that Ω^1 is a finitely generated, projective R–module. (J. Lipman, [**L₁**], (4.6.4) and private communication.)

Class IV contains all pairs $(S/R, \Omega)$, where S/R is a finite locally complete intersection ([**KD**], app. C), R is a noetherian ring and Ω is a differential algebra of R. (E. Kunz, [**KD**], §16.)

Trace maps also can be constructed in some other cases (c.f. (4.17)), however E. Kunz and A. Kliegl gave an example of an integral domain R and a finite free algebra S/R such that with $K := Q(R)$ and $L := K \otimes_R S$ the algebra L/K is a complete intersection, but such that there exists no map $\Omega_{S/\mathbf{Z}}^{\cdot} \to \Omega_{R/\mathbf{Z}}^{\cdot}$ such that the following diagram commutes:

$$
\begin{array}{ccc}
\Omega_{S/\mathbf{Z}}^{\cdot} & \longrightarrow & \Omega_{L/\mathbf{Z}}^{\cdot} \\
\downarrow & & \sigma_{L/K}^{\Omega_{K/\mathbf{Z}}^{\cdot}} \downarrow \\
\Omega_{R/\mathbf{Z}}^{\cdot} & \longrightarrow & \Omega_{K/\mathbf{Z}}^{\cdot}.
\end{array}
$$

Since the system of trace maps for finite locally complete intersections is uniquely determined by the trace axioms TR 1–TR 4 ([**KD**], (16.1)), it follows that there exists no system of traces for all finite projective algebras S/R and arbitrary differential algebras Ω of R satisfying the trace axioms TR 1–TR 4 ([**KD**], §16, ex. 2).

Lipman ([**L₁**], §4) however showed: If the algebra of differential forms of an algebra is replaced by the Hochschild homology of this algebra, then there always exists a canonical trace map in Hochschild homology and the trace of differential forms for the classes I–III can be derived from the trace in Hochschild homology. It will be shown in this paper that the trace maps of Hochschild homology modules satisfy axioms similar to TR 1–TR 8. From this the corresponding properties for traces of differential forms can be derived easily in most cases.

Consequently this paper starts with studying the Hochschild homology functor

$$H. : A \to ADG$$

from the category A of commutative algebras to the category ADG of anti–commutative DG–algebras. It turns out to be useful to extend this functor to the category Topal of topological algebras. This will be done in §1.

In the second section some relations between differential algebras and Hochschild homology modules will be established. It is well known that for every commutative algebra R/k there exists a functorial homomorphism

$$\theta_{R/k}^{\cdot} : \Omega_{R/k}^{\cdot} \to H.(R/k)$$

of anti–commutative DG–algebras ([**L₁**], (1.10.2)). Extending the functor "universal differential algebra"

$$\Omega^{\cdot} : A \to ADG$$

in a suitable way to the category Topal, the above transformation of functors can be extended to the full subcategory Topal~ of all objects of Topal for which the differential algebra $\hat{R}\left[d\hat{R}\right] \subseteq H.(R/k,\tau)$ is complete (in the sense of definition (2.1)). Here $(R/k,\tau)$ denotes an object of Topal, that is a commutative algebra R/k together with a linear topology τ on R.

If in addition $\Omega^n_{(R/k,\tau)}$ is complete as $\hat{R}$–module for all n, then there exists a functorial homomorphism of graded $\hat{R}$–modules

$$\overline{\delta}_{(R/k,\tau)} : H.(R/k,\tau) \to \Omega^{\cdot}_{(R/k,\tau)}$$

satisfying

$$\overline{\delta}^n_{(R/k,\tau)} \circ \theta^n_{(R/k,\tau)} = n!\,id_{\Omega^n_{(R/k,\tau)}}.$$

If R is a $\mathbb{Q}$–algebra or if $\Omega^1_{R/k}$ is a finite free R–module then a left inverse $\delta^{\cdot}_{(R/k,\tau)}$ of $\theta^{\cdot}_{(R/k,\tau)}$ satisfying $n!\delta^n_{(R/k,\tau)} = \overline{\delta}^n_{(R/k,\tau)}$ for all $n \in \mathbb{N}$ can be constructed explicitly.

In §3 a trace for the Hochschild homology of topological algebras will be constructed and the following theorem will be proved:

THEOREM. Let X be the class of all pairs $((S,\tau')/(R,\tau),k)$ with the following properties:

i. k is a noetherian ring and $(R,\tau)/k$ is a topological algebra.

ii. S/R is an algebra and τ' is the linear topology on S induced by τ.

iii. The τ'–adic completion $\left(\hat{S},\widehat{\tau'}\right)$ of (S,τ') is a finite and free module over the τ–adic completion $\left(\hat{R},\hat{\tau}\right)$ of (R,τ).

iv. The topology $\widehat{\tau'}$ on $\hat{S}$ is the linear topology induced by $\hat{\tau}$.

Then there exists a system of canonical morphisms

$$tr_{(S/R,\tau)} : H.(S/k,\tau') \to H.(S/k,\tau),((S,\tau')/(R,\tau),k) \text{ in } X,$$

satisfying the axioms TR 1–TR 8 formulated in terms of Hochschild homology.

In §4 these results will be used to construct the pretrace of Angéniol ([A], (7.1.2)) for pairs $(S/R,\Omega)$, where S/R is a finite projective algebra and Ω is a differential algebra of R. Then the trace maps for differential forms will be constructed for the classes I, II and III and it will be shown that they satisfy the trace axioms. However to prove TR 5 and TR 6 for the classes I and III, I need some additional assumptions. Furthermore it will be shown that the system of trace maps for class II is uniquely determined by the axioms TR 1–TR 8. Considering only the subclasses of the classes I and III consisting of those pairs $(S/R, \Omega)$ for which R is reduced and noetherian and Ω is torsionfree then a uniqueness theorem for their traces can also be proved. For the general case the question of uniqueness is still open.

It is not known whether the trace maps for finite locally complete intersections (class IV) can be defined in a similar way via Hochschild homology. Nevertheless these traces are closely related to the traces of Hochschild homology modules as is shown by the main result of section 5:

If R/k is a local noetherian algebra and if S/R is a finite complete intersection then the following diagram commutes:

$$
\begin{array}{ccc}
\Omega_{S/k}^{\cdot} & \xrightarrow{\ \theta_{S/k}\ } & H.(S/k) \\
{\scriptstyle \sigma_{S/R}^{\Omega_{R/k}^{\cdot}}} \downarrow & & \downarrow {\scriptstyle tr_{S/R}} \\
\Omega_{R/k}^{\cdot} & \xrightarrow{\ \theta_{R/k}\ } & H.(R/k).
\end{array}
$$

The proof of this theorem heavily depends on the Hochschild homology of topological algebras and its properties as derived in §1 and §2.

Using this theorem it can be shown that the various definitions of trace coincide on the intersection of each two of the classes I–IV. Furthermore the trace axioms TR 5 and TR 6 for the classes I and III can be shown in some special cases, and the assumption "2 is not a zero–divisor on R" in the definition of class III can be eliminated in some cases.

The results of section 5 are of interest in particular in the theory of residues of differential forms as developed by Lipman in $[L_1]$. Theorem (5.1) for instance implies that "trace formula II" as stated in $[L_1]$, (4.7.3) holds for local complete intersections. It is not known whether the traces defined in §4 satisfy this formula, however in §6 we will extend Lipman's definition of residues to topological Hochschild homology and universally finite differential forms, and we will use this theory to derive under suitable reducedness assumptions a slightly weaker version of the trace formula:

Let R/k be an algebra and assume that k is noetherian and reduced. Let $f_1, \ldots, f_n \in R$ be an R–regular sequence such that $R/(f_1, \ldots, f_n)$ is finite and flat as a k–module, and

7

assume that S/R is a finite and projective algebra for which a trace map

$$\sigma_{S/R} : \Omega_{S/k}^{\cdot} \to \Omega_{R/k}^{\cdot}$$

in the sense of §4 is defined. Then for $\omega \in \Omega_{S/k}^n$

$$\text{Res}_{S/k}^n \begin{bmatrix} \omega \\ f_1, \ldots, f_n \end{bmatrix} = \text{Res}_{R/k}^n \begin{bmatrix} \sigma_{S/R}(\omega) \\ f_1, \ldots, f_n \end{bmatrix}$$

This formula we will use in section 7 to deduce residue axiom (R4) "transitivity" ([RD], p. 199) for Lipman's residue symbol. This is the only residue formula in [RD] not proved in [L₁].

Acknowledgements:

Sections 1 to 5 are based on the author's dissertation [Hü] done at the University of Regensburg under the supervision of Prof. Dr. E. Kunz. Parts of §1 and §5 and the sections 6 and 7 contain results of research done at Purdue University. I wish to express my gratitude to E. Kunz and J. Lipman for helpful discussions and suggestions. Furthermore I am thankful to Judy Mitchell who typeset the manuscript.

§1. The Hochschild homology and the Hochschild cohomology of a topological algebra.

In this section the definition of Hochschild homology and Hochschild cohomology will be extended to topological algebras. Before doing so we will fix our notation and recall some well–known facts about Hochschild homology and complete tensor products.

A *commutative algebra* is a triple (R, k, ρ), consisting of commutative rings R and k with 1 together with a unitary ring homomorphism $\rho : k \to R$. A morphism $(R, k, \rho) \to (R', k', \rho')$ of commutative algebras is a pair of unitary ring homomorphisms $\psi : k \to k'$ and $\varphi : R \to R'$ such that the following diagram commutes:

$$
\begin{array}{ccc}
R & \xrightarrow{\ \varphi\ } & R' \\
\rho \uparrow & & \uparrow \rho' \\
k & \xrightarrow{\ \psi\ } & k'
\end{array}
$$

This way we get a category, the category A of commutative algebras.

If no confusion is likely an algebra (R, k, ρ) will be denoted by R/k and a morphism $(R, k, \rho) \xrightarrow{(\varphi, \psi)} (R', k', \rho')$ will be denoted by $\varphi : R \to R'$ resp. $\varphi : R/k \to R'/k'$.

A *DG–algebra* $(\Omega, d)/k$ over a commutative ring k is an associative, positively graded k–algebra Ω with 1 together with a morphism $d : \Omega \to \Omega$ of k–modules, satisfying

i. d is homogeneous of degree 1.

ii. $d^2 = 0$.

iii. $d(\omega_1 \omega_2) = d(\omega_1) \cdot \omega_2 + (-1)^{\deg(\omega_1)} \omega_1 d(\omega_2)$ for $\omega_1, \omega_2 \in \Omega$ homogeneous.

A morphism $(\Omega, d)/k \to (\Omega', d)/k'$ of DG–algebras consists of unitary ring homomorphisms $\psi : k \to k'$ and $\varphi : \Omega \to \Omega'$ satisfying:

i. φ is homogeneous of degree 0.

ii. $\varphi \circ d = d \circ \varphi$.

iii. The following diagram commutes:

$$
\begin{array}{ccc}
\Omega & \xrightarrow{\ \varphi\ } & \Omega' \\
\rho \uparrow & & \uparrow \rho' \\
k & \xrightarrow{\ \psi\ } & k'.
\end{array}
$$

With these definitions the DG–algebras form a category which will be denoted by DG. By ADG we denote the full subcategory of DG whose objects are the anti–commutative DG–algebras.

1.1. Remark

i. If $(\Omega, d)/k$ is a DG–algebra, and if $\Omega = \underset{n \in \mathbb{N}}{\oplus} \Omega^n$ is the decomposition of Ω in its homogeneous components, then $1 \in \Omega^0$.

ii. If in addition Ω is anti–commutative, then Ω^0 is a commutative ring with 1 and contained in the center of Ω.

iii. The category $\mathcal{D}$ of differential algebras in the sense of ([KD], (2.1)) is canonically isomorphic to the full subcategory of ADG whose objects $(\Omega, d)/k$ satisfy:

a) $\Omega = \Omega^0 \left[d\Omega^0 \right]$, i.e. as an Ω^0–algebra Ω is generated by its elements of the form $d\omega, \omega \in \Omega^0$.

b) $d\omega \cdot d\omega = 0$ for all $\omega \in \Omega^0$.

Ω is a differential algebra of Ω^0.

iv. Conversely let $(\Omega, d)/k$ be an object of ADG such that $d\omega \cdot d\omega = 0$ for all $\omega \in \Omega^0$. Then the subalgebra $\Omega^0 \left[d\Omega^0 \right]$ of Ω is a differential algebra of Ω^0.

Let R/k be an object of A. By $R^e := R \otimes_k R$ we denote the enveloping algebra of R/k. Then R is an R^e–module via

$$\mu : R \otimes_k R \to R, \mu(r_1 \otimes r_2) = r_1 \cdot r_2,$$

and therefore the nth relative torsion module

$$H_n(R/k) := Tor_n^{(R^e, k)}(R, R)$$

of R with respect to (R^e, k) is well–defined for every $n \in \mathbb{N}$ ([Ho]).

Definition: $H_n(R/k)$ is called the nth *Hochschild homology group* of R/k.
$$H.(R/k) := \underset{n \in \mathbb{N}}{\oplus} H_n(R/k)$$

$H.(R/k)$ can be described as the homology of the Hochschild complex $(\gamma.(R/k), d.)$. This complex is defined as follows:

$\gamma_n(R/k) := T_k^{n+1}(R) := R \otimes_k \cdots \otimes_k R$ is the $(n+1)$–fold tensor product of $R/k (n \in \mathbb{N})$.

$d_n : \gamma_n(R/k) \to \gamma_{n-1}(R/k)$ is given by

$$d_n\,(r_0 \otimes r_1 \otimes \cdots \otimes r_n)$$
$$= \sum_{i=0}^{n-1}(-1)^i r_0 \otimes r_1 \otimes \cdots \otimes r_i r_{i+1} \otimes \cdots \otimes r_n + (-1)^n r_n r_0 \otimes r_1 \otimes \cdots \otimes r_{n-1}$$

for $r_0, \cdots, r_n \in R, n \in \mathbf{N}$ ([**ML**], X. §4).

Multiplication in the first component makes $\gamma_n(R/k)$ an R–module and d_n a homomorphism of R–modules. Therefore $H.(R/k)$ comes equipped with a canonical R–module structure. Since $d_1 = 0$ and $d_0 = 0$ it holds

$$H_0(R/k) = \gamma_0(R/k) = R.$$

If $\varphi : R/k \to R'/k'$ is a morphism in A, then φ defines in a canonical way a morphism $\gamma.(R/k) \to \gamma.(R'/k')$ of complexes. The map induced in homology will be denoted by $H.(\varphi)$.

1.2. THEOREM ([**HKR**], 5.; [**R**], 10.). *H. defines a covariant functor*

$$H. : A \to ADG$$

1.3. *Remark:*

i. Using the Hochschild complex the multiplication on $H.(R/k)$ can be described as follows ([**HKR**], pp. 392–393):

For $r_0 \otimes r_1 \otimes \cdots \otimes r_n \in \gamma_n(R/k)$ and $\tilde{r}_0 \otimes r_{n+1} \otimes \cdots \otimes r_{n+m} \in \gamma_m(R/k)$ set:

$$(r_0 \otimes r_1 \otimes \cdots \otimes r_n) \cdot \left(\tilde{r}_0 \otimes r_{n+1} \otimes \cdots \otimes r_{n+m}\right)$$
$$:= \sum_{\sigma \in G_{n,m}} \operatorname{sign}(\sigma) r_0 \tilde{r}_0 \otimes r_{\sigma(1)} \otimes \cdots \otimes r_{\sigma(n+m)},$$

where $G_{n,m} := \{\sigma \in S_{n+m} : \sigma^{-1}(1) < \cdots < \sigma^{-1}(n), \sigma^{-1}(n+1) < \cdots < \sigma^{-1}(n+m)\}$

(the "shuffle product" of $r_0 \otimes r_1 \otimes \cdots \otimes r_n$ and $\tilde{r}_0 \otimes r_{n+1} \otimes \cdots \otimes r_{n+m}$).

This definition makes $\gamma.(R/k)$ a graded anti-commutative R–algebra satisfying

$$d_{n+m}(x \cdot y) = d_n(x) \cdot y + (-1)^n x \cdot d_m(y) \text{ for } x \in \gamma_n(R/k), y \in \gamma_m(R/k).$$

Furthermore $x^2 = 0$ for all $x \in \gamma_1(R/k)$. Consequently this multiplication induces a graded anti-commutative R–algebra structure on $H.(R/k)$, satisfying $x^2 = 0$ for all $x \in H_1(R/k)$.

ii. The differentiation d_R on $H.(R/k)$ is induced by the mapping

$$\tilde{\delta} : \gamma.(R/k) \to \gamma.(R/k)$$

given by

$$\widetilde{\delta}\left(r_0 \otimes r_1 \otimes \cdots \otimes r_n\right) = \sum_{i=0}^{n} (-1)^{in} 1 \otimes r_i \otimes \cdots \otimes r_n \otimes r_0 \otimes \cdots \otimes r_{i-1} \text{ for } r_0, r_1, \cdots, r_n \in R.$$

This morphism is k–linear and homogeneous of degree 1. Furthermore this map satisfies $\widetilde{\delta}_{n-1} \circ d_n + d_{n+1} \circ \widetilde{\delta}_n = 0$ for all $n \in \mathbf{N}$, and therefore induces a map in homology. In order to prove the Leibniz–formula let

$$h_{n,m} : \gamma_n(R/k) \times \gamma_m(R/k) \to \gamma_{n+m+2}(R/k) \quad (n, m \in \mathbf{N})$$

be the k–bilinear mapping given by:

$$h_{n,m}\left(r_1 \otimes r_2 \otimes \cdots \otimes r_{n+1}, r_{n+2} \otimes r_{n+3} \otimes \cdots \otimes r_{n+m+2}\right)$$
$$= \sum_{\sigma \in K_{n,m}} \text{sign}\,(\sigma) 1 \otimes r_{\sigma(1)} \otimes \cdots \otimes r_{\sigma(n+m+2)},$$

where $K_{n,m}$ is defined as follows:

Let $\tau_0 := \langle 1, 2, \cdots, n+1 \rangle \in S_{n+m+2}$ and let $\tau_1 := \langle n+2, n+3, \cdots, n+m+2 \rangle \in S_{n+m+2}$ and set

$$K_{n,m} := \big\{ \sigma \in S_{n+m+2} : \sigma^{-1}(1) < \sigma^{-1}(n+2),$$

and there exist $p, q \in \mathbf{N}$ such that

$$\sigma^{-1}\left(\tau_0^p(1)\right) < \sigma^{-1}\left(\tau_0^p(2)\right) < \cdots < \sigma^{-1}\left(\tau_0^p(n+1)\right) \text{ and}$$
$$\sigma^{-1}\left(\tau_1^q(n+2)\right) < \sigma^{-1}\left(\tau_1^q(n+3)\right) < \cdots < \sigma^{-1}\left(\tau_1^q(n+m+2)\right) \big\}.$$

A tedious but straightforward calculation shows:

It holds for $a \in \gamma_n(R/k)$ and $b \in \gamma_m(R/k)$:

$$d_{n+m+2}\left(h_{n,m}(a, b)\right) - (-1)^n h_{n,m-1}\left(a, d_m(b)\right) + h_{n-1,m}\left(d_n(a), b\right)$$
$$= (-1)^{n+1} \widetilde{\delta}_{n+m}(a \cdot b) + (-1)^n \widetilde{\delta}_n(a) \cdot b + a \cdot \widetilde{\delta}_m(b).$$

This implies the Leibniz–formula:

$$d_R(x \cdot y) = d_R(x) \cdot y + (-1)^{\deg(x)} x \cdot d_R(y) \text{ for } x, y \in H.(R/k) \text{ homogeneous.}$$

Furthermore it can be shown that $\widetilde{\delta}$ is homotopic to $(id - p.) \circ \widetilde{\delta}$, where $p.$ is defined by

$$p_n\left(r_0 \otimes r_1 \otimes \cdots \otimes r_n\right) = (-1)^n r_n \otimes r_0 \otimes \cdots \otimes r_{n-1}$$
$$\left(r_0, r_1, \cdots, r_n \in R, n \in \mathbf{N}\right).$$

Since $\left((id - p.) \circ \widetilde{\delta}\right)^2 = 0$, this implies:

$$d_R^2 = 0.$$

(For details see: [**R**], pp. 216–222.)

    iii. $H.(R/k)$ is an anti–commutative DG–algebra over k satisfying:

    a) $H_0(R/k) = R$.

    b) $x^2 = 0$ for every $x \in H_1(R/k)$.

However in general $H.(R/k)$ is not a differential algebra of R/k as the later example (1.16) will show.

    In the remaining part of this section the definition of the functor $H.$ will be extended to the category of topological algebras.

    A *topological algebra* is a triple $((R, \tau), (k, \tau_0), \rho)$ where (R, k, ρ) is a commutative algebra, τ is a linear topology on R, and τ_0 is a linear topology on k such that ρ is continuous with respect to these topologies ([**EGA O$_1$**], §7).

    A morphism $((R, \tau), (k, \tau_0), \rho) \rightarrow ((R', \tau'), (k', \tau_0'), \rho')$ of topological algebras is a morphism $R/k \xrightarrow{(\varphi, \psi)} R'/k'$ of commutative algebras such that φ and ψ are continuous in the implied topologies.

    These definitions make the topological algebras a category which will be denoted by Topal. By Topal$_{(k, \tau_0)}$ we denote the subcategory of Topal whose objects are the topological (k, τ_0)–algebras and whose morphisms act as identity on (k, τ_0). Topal^ denotes the full subcategory of Topal whose objects $((R, \tau), (k, \tau_0), \rho)$ satisfy: R is complete with respect to its topology τ. Here "complete" always means "complete and separated" in the sense of [**B$_2$**].

    If no confusion is likely a topological algebra $((R, \tau), (k, \tau_0), \rho)$ will be denoted by $(R, \tau)/(k, \tau_0)$ and a morphism

$$((R, \tau), (k, \tau_0), \rho) \xrightarrow{(\varphi, \psi)} ((R', \tau'), (k', \tau_0'), \rho')$$

will be denoted by $\varphi : (R, \tau) \rightarrow (R', \tau')$ resp. by $\varphi : (R, \tau)/(k, \tau_0) \rightarrow (R', \tau')/(k', \tau_0')$.

    The category A of commutative algebras is canonically isomorphic to the full subcategory of Topal^ whose objects $((R, \tau), (k, \tau), \rho)$ satisfy: Both τ and τ_0 are the discrete topologies. In this case the symbols τ and τ_0 will be omitted.

    Next let us recall completions and complete tensorproducts of topological algebras.

    A *completion* of a topological ring (R, τ) is a topological ring $\left(\hat{R}, \hat{\tau}\right)$ together with a continuous morphism $(R, \tau) \xrightarrow{i} \left(\hat{R}, \hat{\tau}\right)$ of rings satisfying:

i. $\hat{R}$ is complete and separated in its topology $\hat{\tau}$.

ii. If (S, τ') is a complete and separated ring and if $\varphi : (R, \tau) \to (S, \tau')$ is a continuous morphism of rings there exists a unique continuous morphism $\hat{\varphi} : \left(\hat{R}, \hat{\tau}\right) \to (S, \tau')$ of rings such that $\varphi = \hat{\varphi} \circ i$.

1.4. *Remark*: Every topological ring (R, τ) has a completion $\left(\hat{R}, \hat{\tau}\right)$, unique up to canonical isomorphism.

If $\mathfrak{U}$ is a fundamental set of neighborhoods of $0 \in R$, consisting of ideals of R, then we have $\hat{R} = \varprojlim_{U \in \mathfrak{U}} R/U$ together with the limit topology $\hat{\tau}$, and $i : R \to \hat{R}$ is the canonical mapping $R \to \varprojlim_{U \in \mathfrak{U}} R/U$.

The topological closures $\overline{U}$ of the images of the ideals $U \in \mathfrak{U}$ in $\hat{R}$ form a fundamental set of neighborhoods of $0 \in \hat{R}$ in the topology $\hat{\tau}$. Each $\overline{U}$ is an ideal of $\hat{R}$ and $\hat{R}/\overline{U} \cong R/U$.

Completion defines a covariant functor

$$\hat{} : \text{Topal} \to \text{Topal}\hat{}$$

mapping an object $(R, \tau)/(k, \tau_0)$ to $\left(\hat{R}, \hat{\tau}\right)/(k, \tau_0)$.

Proofs of the above facts can be found in ([**B₂**].III), in particular in §3.4, §6.4, §6.5 and §7.3.

Given two topological (k, τ_0)–algebras (R, τ) and (S, τ') their tensor product $R \otimes_k S$ comes equipped with a canonical topology, the tensor product topology $\tau \otimes \tau'$, defined as follows ([**EGA O₁**], (7.7); [**B₁**], III §2, ex. 28; [**S**], V. B. 2):

Given fundamental sets of neighborhoods $\mathfrak{U}$ resp. $\mathfrak{V}$ of 0 in R resp. S, consisting of ideals of R resp. S, the ideals $(im\,(U \otimes_k S) + im\,(R \otimes_k V)) \subseteq R \otimes_k S, U \in \mathfrak{U}, V \in \mathfrak{V}$ form a fundamental set of neighborhoods of 0 in $R \otimes_k S$ for the topology $\tau \otimes \tau'$.

Clearly $(R \otimes_k S, \tau \otimes \tau')$ is a topological (k, τ_0)–algebra. Its completion is denoted by $\left(R\widehat{\otimes}_k S, \tau \hat{\otimes} \tau'\right)/(k, \tau_0)$ and is called the *complete tensor product* of (R, τ) and (S, τ') over (k, τ_0).

1.5. *Remark*: Let (R, τ) and (S, τ') be topological (k, τ_0)–algebras.

i. $R\widehat{\otimes}_k S$ doesn't depend on the topology τ_0 of k.

ii. Given fundamental sets of neighborhoods $\mathfrak{U}$ resp. $\mathfrak{V}$ of 0 in R resp. S, consisting of ideals of R resp. S, there exists a canonical isomorphism

$$R\widehat{\otimes}_k S \cong \varprojlim_{U \in \mathfrak{U}, V \in \mathfrak{V}} R/U \otimes_k S/V \qquad ([\mathbf{B_1}], \text{III. §2, ex.28a}).$$

In particular it holds

$$R\widehat{\otimes}_k R \cong \varprojlim_{U \in \mathfrak{U}} R/U \otimes_k R/U \qquad \text{([S], V.B.2.c).}$$

iii. There exists a canonical isomorphism of topological (k, τ_0)–algebras

$$\left(R\widehat{\otimes}_k S, \tau\hat{\otimes}\tau'\right) \cong \left(\hat{R}\widehat{\otimes}_k\hat{S}, \hat{\tau}\hat{\otimes}\hat{\tau'}\right) \qquad \text{([B}_1\text{)], III. §2, ex. 28a).}$$

iv. Let (T, τ'') be a topological (R, τ)–algebra and suppose that τ'' is the linear topology on T induced by τ. Then the canonical isomorphism

$$T \otimes_R (R \otimes_k S) \rightarrow T \otimes_k S$$

is a homeomorphism and induces a topological isomorphism

$$T\widehat{\otimes}_R (R\widehat{\otimes}_k S) \rightarrow T\widehat{\otimes}_k S \qquad \text{([EGA O}_1\text{], (7.7.8)).}$$

v. If τ is the $\mathfrak{I}$–adic topology on R for some ideal $\mathfrak{I} \subseteq R$ and if τ' is the $\mathfrak{J}$–adic topology on S for some ideal $\mathfrak{J} \subseteq S$, then $\mathfrak{K} := (im\,(\mathfrak{I} \otimes_k S) + im\,(R \otimes_k \mathfrak{J})) \subseteq R \otimes_k S$ is an ideal, and $R\widehat{\otimes}_k S$ is the $\mathfrak{K}$–adic completion of $R \otimes_k S$ ([EGA O$_1$], (7.7.7)).
If $\mathfrak{I}$ and $\mathfrak{J}$ are finitely generated, then the topology on $R\widehat{\otimes}_k S$ is the $\mathfrak{K} \cdot R\widehat{\otimes}_k S$–adic topology ([GS], (4.3); [N] (17.4)).

vi. Suppose that τ_0 is the $\mathfrak{I}$–adic topology on k for some ideal $\mathfrak{I} \subseteq k$, that (S, τ') is a finitely generated k–module, and that τ' is the $\mathfrak{I}S$–adic topology on S. Furthermore suppose that τ is the $\mathfrak{J}$–adic topology on R for some ideal $\mathfrak{J}$, and that R is complete with respect to τ. Then the canonical morphism

$$R \otimes_k S \rightarrow R\widehat{\otimes}_k S$$

is an isomorphism ([EGA O$_1$], (7.7.9)).

Further properties of complete tensor products follow from its universal property ([EGA O$_1$], (7.7.6)).

1.6. THEOREM. $\hat{\otimes}k$ is the fibered sum in the category $\mathrm{Topal}_{(k,\tau_0)}$.
More precisely it holds:

Given topological (k,τ_0)–algebras (R,τ) and (S,τ') we have that the canonical homomorphisms $(R,\tau) \to (R \otimes_k S, \tau \otimes \tau')$ and $(S,\tau') \to (R \otimes_k S, \tau \otimes \tau')$ are continuous and induce mappings $\rho : R \to R\widehat{\otimes}_k S$ and $\sigma : S \to R\widehat{\otimes}_k S$ satisfying:

If (T,τ'') is a complete topological (k,τ_0)–algebra and if $u : R \to T$ and $v : S \to T$ are morphisms of topological (k,τ_0)–algebras, then there exists a unique continuous morphism $w : R\widehat{\otimes}_k S \to T$ of topological (k,τ_0)–algebras such that $w \circ \rho = u$ and $w \circ \sigma = v$.

In particular if $(R,\tau) = (S,\tau')$ and if $i : R \to \hat{R}$ is the canonical mapping, then there exists a continuous homomorphism $\mu : R\widehat{\otimes}_k R \to \hat{R}$ of k–algebras such that $\mu \circ \sigma = i = \mu \circ \rho$.

1.7. COROLLARY. Given topological (k,τ_0)–algebras $(R,\tau), (S,\tau')$ and (T,τ'') there exist canonical isomorphisms of topological (k,τ_0)–algebras

$$\left(R\widehat{\otimes}_k S, \tau\hat{\otimes}\tau'\right) \xrightarrow{\sim} \left(S\widehat{\otimes}_k R, \tau'\hat{\otimes}\tau\right)$$

and

$$\left((R\widehat{\otimes}_k S)\,\widehat{\otimes}_k T, (\tau\hat{\otimes}\tau')\,\hat{\otimes}\tau''\right) \xrightarrow{\sim} \left(R\widehat{\otimes}_k (S\widehat{\otimes}_k T), \tau\hat{\otimes}(\tau'\hat{\otimes}\tau'')\right).$$

Therefore we will write $\left(R\widehat{\otimes}_k S\widehat{\otimes}_k T, \tau\hat{\otimes}\tau'\hat{\otimes}\tau''\right)$ instead of $\left((R\hat{\otimes}_k S)\,\hat{\otimes}_k T, (\tau\hat{\otimes}\tau')\,\hat{\otimes}\tau''\right)$ in the sequel.

In order to generalize the Hochschild homology functor to the category of topological algebras we need some information about n–fold complete tensor products of topological algebras $(R,\tau)/(k,\tau_0)$ which will be denoted by $T_k^n(R,\tau)(n \in \mathbf{N})$.

1.8. Remark: Let $\mathfrak{I} \subseteq R$ be an ideal of R, τ the $\mathfrak{I}$–adic topology on R and denote by $\mathfrak{I}_{(n)}$ the image of the canonical homomorphism

$$\mathfrak{I} \otimes_k R \otimes_k \cdots \otimes_k R \oplus R \otimes_k \mathfrak{I} \otimes_k \cdots \otimes_k R \oplus \cdots \oplus R \otimes_k R \otimes_k \cdots \otimes_k \mathfrak{I} \to T_k^n(R).$$

Then it holds:

i. $T_k^n(R,\tau)$ is the $\mathfrak{I}_{(n)}$–adic completion of $T_k^n(R)$ ((1.5)v.,).

ii. $\mathfrak{I}_{(n)} T_k^n(R,\tau) \subseteq \bar{\mathfrak{I}}_{(n)} \subseteq Rad(T_k^n(R,\tau))$, where $\bar{\mathfrak{I}}_{(n)}$ denotes the topological closure of the image of $\mathfrak{I}_{(n)}$ in $T_k^n(R,\tau)$.

iii. If $T_k^n(R)$ is noetherian then $T_k^n(R, \tau)$ is noetherian (by i. and [**AM**], (10.26)).

In the following we always will assume that τ_0 is the discrete topology on k, and we will write $(R, \tau)/k$ instead of $(R, \tau)/(k, \tau_0)$. In this situation $\left(\widehat{R^e}, \widehat{\tau^e}\right) := \left(R \widehat{\otimes}_k R, \tau \widehat{\otimes} \tau\right)$ is called the *enveloping algebra* of the topological algebra $(R, \tau)/k$.

Let $(R, \tau)/k$ be a topological algebra. The *topological Hochschild complex* $(\gamma.(R/k, \tau), d.)$ is defined to be the projective limit, formed componentwise, of the projective system of Hochschild complexes $(\gamma.(R/U/k), d_U.)$, where U runs through the open ideals of R, i.e.

$$\gamma_n(R/k, \tau) = \varprojlim_U \gamma_n(R/U/k),$$

$$d_n = \varprojlim_U d_{n,U} \text{ for all } n \in \mathbf{Z}.$$

Clearly $(\gamma.(R/k, \tau), d.)$ is a complex of $\hat{R}$–modules and $\gamma_n(R/k, \tau) = T_k^{n+1}(R, \tau)$ for all $n \in \mathbf{N}$.

1.9. Definition: $H_n(R/k, \tau) := H_n((\gamma.(R/k, \tau), d.))$ is called the nth topological Hochschild homology group of the topological algebra $(R, \tau)/k$.

$$H.(R/k, \tau) := \bigoplus_{n \in \mathbf{N}} H_n(R/k, \tau).$$

1.10. Remark:

i. $H.(R/k, \tau)$ is a graded $\hat{R}$–module and $H_0(R/k, \tau) = \hat{R}$.

ii. If τ is the discrete topology on R, then it holds:

$$H.(R/k, \tau) = H.(R/k)$$

iii. Let $\varphi : (R, \tau)/k \to (R', \tau')/k'$ be a morphism of topological algebras. By (1.6) φ induces a canonical homomorphism $\gamma.(R/k, \tau) \to \gamma.(R'/k', \tau')$ of complexes. The map induced in homology will be denoted by $H.(\varphi)$.

In particular there exists a canonical mapping

$$H.(R/k) \to H.(R/k, \tau).$$

iv. Let $\left(\hat{R}, \hat{\tau}\right)/k$ be the completion of $(R, \tau)/k$ and let $i : R \to \hat{R}$ be the canonical morphism. Then the mapping

$$H.(i) : H.(R/k, \tau) \to H.\left(\hat{R}/k, \hat{\tau}\right)$$

is an isomorphism of graded $\hat{R}$–modules.

Like the classical Hochschild homology, $H.(R/k, \tau)$ comes equipped with a canonical anti-commutative DG–algebra structure. More precisely:

1.11. THEOREM. H. *defines a covariant functor*

$$H. : Topal \rightarrow ADG$$

whose restriction to the subcategory A of commutative algebras is the classical Hochschild homology functor of theorem (1.2).

Proof: Clearly H. defines a covariant functor

$$H. : Topal \rightarrow (abelian groups).$$

i. The shuffle product defined in (1.3)i, induces a projective system of R/U–linear maps

$$\gamma_n(R/U/k) \otimes_R \gamma_m(R/U/k) \rightarrow \gamma_{n+m}(R/U/k) \quad (n, m \in \mathbf{Z})$$

where U runs through the open ideals of R. Passing to the projective limit we get well–defined $\hat{R}$–bilinear morphisms

$$\gamma_n(R/k, \tau) \times \gamma_m(R/k, \tau) \rightarrow \gamma_n(R/k, \tau) \widehat{\otimes}_R \gamma_m(R/k, \tau)$$
$$\rightarrow \gamma_{n+m}(R/k, \tau) \quad (n, m \in \mathbf{Z})$$

which are continuous in both factors, and which define a multiplication on $\gamma.(R/k, \tau)$. Clearly this definition makes

$$\gamma.(R/k) \rightarrow \gamma.(R/k, \tau)$$

a morphism of graded R–algebras.

It is an easy exercise in projective limits to deduce the following from the corresponding statements in (1.3)i:

a) The above definition makes $\gamma.(R/k, \tau)$ an anti–commutative graded $\hat{R}$–algebra.

b) $d_{n+m}(x \cdot y) = d_n(x) \cdot y + (-1)^n x \cdot d_m(y)$ for $x \in \gamma_n(R/k, \tau)$ and $y \in \gamma_m(R/k, \tau)$.

c) $x^2 = 0$ for all $x \in \gamma_1(R/k, \tau)$.

Hence we get an induced graded anti–commutative $\hat{R}$–algebra structure on $H.(R/k, \tau)$ satisfying $x^2 = 0$ for all $x \in H_1(R/k, \tau)$.

ii. There exists a k–linear complex homomorphism

$$\hat{\delta} : \gamma.(R/k,\tau) \to \gamma.(R/k,\tau)$$

which is homogeneous of degree 1, such that the following diagram commutes

$$
\begin{array}{ccc}
\gamma.(R/k) & \xrightarrow{\text{can.}} & \gamma.(R/k,\tau) \\
\Big\downarrow{\tilde{\delta}} & & \Big\downarrow{\hat{\delta}} \quad (*) \\
\gamma.(R/k) & \xrightarrow{\text{can.}} & \gamma.(R/k,\tau),
\end{array}
$$

$\tilde{\delta}$ being the map defined in (1.3)ii.

$\hat{\delta}$ induces a morphism

$$d_R : H.(R/k,\tau) \to H.(R/k,\tau)$$

having the following properties:

a) $d_R \bullet d_R = 0$.

b) $d_R(x \cdot y) = d_R(x) \cdot y + (-1)^{\deg(x)} x \cdot d_R(y)$ for homogeneous $x, y \in H.(R/k,\tau)$.

Similar to the statements in i., these assertions follow from the discrete case:

$\tilde{\delta} : \gamma.(R/k) \to \gamma.(R/k)$ as defined in (1.3)ii, induces a projective system of complex homomorphisms

$$\tilde{\delta}_U : \gamma.(R/U/k) \to \gamma.(R/U/k)$$

where U runs through the open ideals of R. Passing to the projective limit we get a k–linear complex homomorphism

$$\hat{\delta} : \gamma.(R/k,\tau) \to \gamma.(R/k,\tau)$$

homogeneous of degree 1 which makes $(*)$ commutative. Using the auxiliary morphisms defined in (1.3)ii, resp. ([R],10.) one concludes that the map d_R induced in homology has the properties a) and b).

Obviously the constructions in i. and ii. are functorial in $(R,\tau)/k$, thus proving the theorem.

Next we will discuss the relations between Hochschild homology and Tor–functors.

Recall that for an algebra R/k the *unnormalized bar resolution* $(\beta.(R,R),\partial.)$ of R as an R^e–module is defined as follows ([ML], X.2):

$\beta_n(R,R) = T_k^{n+2}(R) = R \otimes_k \cdots \otimes_k R$ is the $(n+2)$-fold tensor product of R/k and
$\partial_n (r_0 \otimes r_1 \otimes \cdots \otimes r_{n+1}) = \sum_{i=0}^{n} (-1)^i r_0 \otimes \cdots \otimes r_i r_{i+1} \otimes \cdots \otimes r_{n+1}$ $(r_0, \cdots, r_{n+1} \in R, n \in \mathbb{N})$.
The $\beta_n(R,R)$ are R^e–modules via

$$(r \otimes s) \cdot (r_0 \otimes r_1 \otimes \cdots \otimes r_n \otimes r_{n+1}) = (rr_0) \otimes r_1 \otimes \cdots \otimes r_n \otimes (r_{n+1}s)$$

$(r, s, r_0, \cdots, r_{n+1} \in R)$, and the ∂_n are R^e–linear with respect to this structure ($n \in \mathbb{N}$).

Defining $\mu : \beta_0(R,R) = R^e \to R$ by $\mu(r_0 \otimes r_1) = r_0 r_1$ it is well known that

$$(\beta_.(R,R), \partial_.) \xrightarrow{\mu} R \to 0$$

is an exact complex of R^e–modules having a k–homotopy s given by $s(r_0 \otimes \cdots \otimes r_n) = 1 \otimes r_0 \otimes \cdots \otimes r_n$ $(r_0, \cdots, r_n \in R)$ ([**ML**], X.2).

If $(R, \tau)/k$ is a topological algebra the complexes $(\beta_.(R/U, R/U), \partial_.)$ ($U \subseteq R$ an open ideal) form a projective system of complexes of $(R/U)^e$–modules. By passing componentwise to its projective limit we get a complex $(\beta_.(R/k, \tau), \partial_.)$ of $\widehat{R^e}$–modules with $\beta_n(R/k, \tau) = T_k^{n+2}(R, \tau)$ for all $n \in \mathbb{N}$ together with a morphism

$$\beta_.(R/k, \tau) \xrightarrow{\mu} \hat{R}.$$

Furthermore s induces a k–homotopy of the complex

$$(\beta_.(R/k, \tau), \partial_.) \xrightarrow{\mu} \hat{R} \to 0$$

so that $(\beta_.(R/k, \tau), \partial_.)$ is an $\widehat{R^e}$–exact resolution of $\hat{R}$ together with a k–homotopy s.

$(\beta_.(R/k, \tau), \partial_.)$ is called *topological bar resolution* of the algebra $\left(\hat{R}, \hat{\tau}\right)/k$.

1.12. Remark:

i. The isomorphism of R^e–modules

$$\varphi : \beta_n(R,R) \to R^e \otimes_k T_k^n(R)$$

given by

$$\varphi(r_0 \otimes r_1 \otimes \cdots \otimes r_n \otimes r_{n+1}) = (r_0 \otimes r_{n+1}) \otimes r_1 \otimes \cdots \otimes r_n \quad (r_0, \cdots, r_{n+1} \in R)$$

is a homeomorphism and therefore induces an $\widehat{R^e}$–isomorphism

$$\beta_n(R/k, \tau) \rightarrow \widehat{R^e} \widehat{\otimes}_k T_k^n(R, \tau)$$

ii. Suppose that either τ is the discrete topology on R or that R is noetherian and $\mathfrak{I} \subseteq R$ is an ideal such that τ is the $\mathfrak{I}$-adic topology on R, and such that $T_k^n(R, \tau)$ is noetherian for every $n \in \mathbb{N}$. Then there exist canonical isomorphisms of complexes of $\widehat{R^e}$–modules

$$\left(\hat{R} \otimes_{\widehat{R^e}} \beta.(R/k, \tau), id \otimes \partial. \right)$$
$$\cong \left(\hat{R} \hat{\otimes}_{\widehat{R^e}} \beta.(R/k, \tau), id \hat{\otimes} \partial. \right)$$
$$\cong (\gamma.(R/k, \tau), d.).$$

If τ is the discrete topology then this is the isomorphism of [**ML**], X.4; otherwise its existence follows from i., (1.5)iv., and (1.5) vi.

1.13. PROPOSITION. *Suppose either that R/k is flat and that τ is the discrete topology on R or that R/k is flat, $T_k^n(R)$ is noetherian for all $n \in \mathbb{N}$ and τ is the $\mathfrak{I}$-adic topology on R for some ideal $\mathfrak{I} \subseteq R$. Then there exist canonical isomorphisms of graded $\hat{R}$–modules*

$$H.(R/k, \tau) \cong \operatorname{Tor}^{\widehat{R^e}}. \left(\hat{R}, \hat{R} \right)$$

Proof: In both cases we have by (1.12)ii.:

$$H_n(R/k, \tau) \cong H_n \left(\left(\hat{R} \otimes_{\widehat{R^e}} \beta.(R/k, \tau), id \otimes \partial. \right) \right)$$

Hence it suffices to show:

$\beta_n(R/k, \tau)$ is a flat $\widehat{R^e}$–module for all $n \in \mathbb{N}$.

If τ is discrete, this is well known ((1.12)ii; [**Ma**], 3.B, 3.C); otherwise it follows from

1.14. LEMMA. *Suppose R/k is flat, $T_k^n(R)$ is noetherian for all $n \in \mathbb{N}$ and τ is the $\mathfrak{I}$-adic topology on R for some ideal $\mathfrak{I} \subseteq R$.*

Then $\widehat{R^e} \widehat{\otimes}_k T_k^n(R, \tau)$ is a flat $\widehat{R^e}$–module for all $n \in \mathbb{N}$.

Proof: $R^e \otimes_k T_k^n(R)/R^e$ is flat since $T_k^n(R)/k$ is flat. $\widehat{R^e} \widehat{\otimes}_k T_k^n(R, \tau) \cong T_k^{n+2}(R, \tau)$ is the $\mathfrak{I}_{(n+2)}$-adic completion of the noetherian ring $T_k^{n+2}(R)$ (with the ideal $\mathfrak{I}_{(n+2)}$ as defined in

(1.8)), hence it is flat over $R^e \otimes_k T^n_k(R)$ by [**AM**], (10.14).

Consequently $\widehat{R^e} \widehat{\otimes}_k T^n_k(R, \tau)$ is a flat R^e–module.

$\widehat{R^e} \widehat{\otimes}_k T^n_k(R, \tau)$ is idealwise separated for $\mathfrak{I}_{(2)} \widehat{R^e}$ since

$$\mathfrak{I}_{(2)} T^{n+2}_k(R, \tau) \subseteq \mathfrak{I}_{(n+2)} T^{n+2}_k(R, \tau) \subseteq \text{ rad } \left(T^{n+2}_k(R, \tau) \right) \quad ((1.8)\text{ii.}).$$

It follows that $T^{n+2}_k(R, \tau)$ is a flat $\widehat{R^e}$–module by [**B$_1$**], III.§5.4, Cor. .

1.15. COROLLARY. *Let R/k be flat and let $N \subseteq R$ be a multiplicatively closed subset. Then the functorial homomorphism*

$$H.(R/k) \to H.(R_N/k)$$

induces a canonical isomorphism

$$H.(R/k)_N \to H.(R_N/k).$$

Proof: Let $N \otimes N := \{n_1 \otimes n_2 \in R^e : n_1, n_2 \in N\}$. Then $N \otimes N \subseteq R^e$ is a multiplicatively closed set. Using [**Ma**], 3.E one gets canonical isomorphisms of $(R_N)^e$–modules:

$$Tor.^{(R_N)^e} (R_N, R_N)$$

$$= Tor.^{(R^e)_{N \otimes N}} (R_{N \otimes N}, R_{N \otimes N})$$

$$= Tor.^{R^e} (R, R)_{N \otimes N}$$

$$= Tor.^{R^e} (R, R)_N.$$

From (1.13) the claim follows.

1.16. *Example:* Let k be a noetherian ring, and let $f = \sum_{n \in \mathbb{N}} a_n X^n \in k[X]$ be a quasi–regular element of $k[X]$, e.g. a monic polynomial. Suppose that $R := k[X]/(f)$ is flat and finitely generated as a k–module, and set $f' := \sum_{n \in \mathbb{N}} n a_n X^{n-1}$.

In this case $R^e = k[X, Y]/(f, g)$ where $g = f(Y) = \sum_{n \in \mathbb{N}} a_n Y^n$ and therefore we have $R = k[X, Y]/(f, Y - X)$ as an R^e–module. Both $\{f, g\}$ and $\{f, Y - X\}$ are $k[X, Y]$–quasi–regular sequences by [**KD**], (C.4) satisfying $(f, g) \subseteq (f, Y - X)$, and it holds:

$$g = a(Y - X) + f \text{ where } a := \sum_{n \in \mathbb{N}} a_n \left(\sum_{i=0}^{n-1} Y^{n-1-i} X^i \right)$$

Denoting by $\overline{h}$ the image of an element $h \in k[X,Y]$ in R^e and by $\overline{\overline{h}}$ its image in R, $\overline{a}$ consequently is the image of a transition determinant Δ from $(f, Y - X)$ to (f, g) in R^e. Since $\overline{\mathfrak{J}} := (f, Y - X)/(f, g) = (\overline{Y - X}) \subseteq R^e$ we conclude from a theorem of Wiebe ([**KD**], (E.21)) that $Ann_{R^e}(\overline{a}) = (\overline{X - Y})$ and $Ann_{R^e}(\overline{X - Y}) = (\overline{a})$. Therefore we get an R^e–free resolution

$$(X., \partial.) \overset{\mu}{\to} R \to 0$$

of R such that

$$X_n = R^e \text{ for all } n \in \mathsf{N},$$
$$\partial_n = \begin{cases} \mu_{\overline{X-Y}} & \text{for } n \in \mathsf{N}_+ \text{ odd} \\ \mu_{\overline{a}} & \text{for } n \in \mathsf{N}_+ \text{ even} \end{cases}$$

with $\mu : X_0 \to R$ being the canonical map.

By (1.13) $H_n(R/k) = H_n(R \otimes_{R^e} X., id \otimes \partial.)$ where

$$R \otimes_{R^e} X_n = R \text{ for all } n \in \mathsf{N},$$
$$id \otimes \partial_n = \begin{cases} \mu_{\overline{\overline{X-Y}}} = 0 & \text{for } n \in \mathsf{N}_+ \text{ odd} \\ \mu_{\overline{\overline{a}}} = \mu_{\overline{\overline{f'}}} & \text{for } n \in \mathsf{N}_+ \text{ even.} \end{cases}$$

Therefore we get canonical isomorphisms of R–modules

$$H_0(R/k) = R$$
$$H_{2n-1}(R/k) \cong R/\left(\overline{\overline{f'}}\right) \qquad \text{for } n \in \mathsf{N}_+$$
$$H_{2n}(R/k) \cong \text{Ann}_R\left(\overline{\overline{f'}}\right) \qquad \text{for } n \in \mathsf{N}_+.$$

In particular it follows that $H.(R/k)$ is a differential algebra of R/k if and only if $f' \in R$ is a unit which in turn is equivalent to R/k being étale ([**KD**], (6.10) and (4.14)).

Immediately from the explicit description of Hochschild homology as the homology of the Hochschild complex we get the following result:

1.17. PROPOSITION. *Let $(R_\lambda/k)_{\lambda \in \Lambda}$ be a direct system of k-algebras, and let $R = \varinjlim_{\lambda \in \Lambda} R_\lambda$. Then there exists a canonical isomorphism of graded R-algebras*

$$\varinjlim_{\lambda \in \Lambda} H.(R_\lambda/k) \to H.(R/k)$$

Proof: Clearly the $(\gamma.(R_\lambda/k), d.)$ are a direct system of complexes of R_λ–modules, and by $[\mathbf{B_3}]$, II.§6.3, Prop. 7:

$$\varinjlim_{\lambda \in \Lambda} (\gamma.(R_\lambda/k), d.) = (\gamma.(R/k), d.).$$

Since $\varinjlim$ is an exact functor ($[\mathbf{B_3}]$, II. §6.2, Prop. 3) we get an isomorphism of graded R–modules

$$\varinjlim_{\lambda \in \Lambda} H.(R_\lambda/k) \to H.(R/k).$$

This isomorphism is precisely the map induced from the canonical homomorphisms $H.(R_\lambda/k) \to H.(R/k)$, hence it is an isomorphism of R–algebras.

From the definition of Hochschild homology as a relative torsion product, the following result about its behaviour with respect to direct products follows:

1.18. PROPOSITION ([ML], X. THM. 6.2). *Let $R = R_1 \times \cdots \times R_n$ be a direct product of k–algebras. Then the homomorphism*

$$H.(R/k) \to H.(R_1/k) \times \cdots \times H.(R_n/k)$$

induced by the canonical projections $R \to R_i$ is an isomorphism of DG–algebras.

Similarly to the theory of Hochschild homology modules of topological algebras a theory of Hochschild cohomology modules can be developed in this category. The Hochschild cohomology groups of R with coefficients in an R–bimodule M, denoted by $H^{\cdot}(R/k, M)$, can be described as the cohomology of the complex $(\bigoplus_{n \in \mathbf{N}} \mathrm{Mult}_k^n(R, M), \delta^{\cdot})$ which is defined as follows:

$\mathrm{Mult}_k^n(R, M)$ denotes the k–multilinear functions on the n–fold cartesian product of R, and the coboundary operator

$$\delta^n : \mathrm{Mult}_k^n(R, M) \to \mathrm{Mult}_k^{n+1}(R, M)$$

is given by the formula

$$\delta^n f(r_1, \ldots, r_{n+1}) = (-1)^{n+1}\{r_1 f(r_2, \ldots, r_{n+1})$$
$$+ \sum_{i=1}^{n}(-1)^i f(r_1, \ldots, r_i r_{i+1}, \ldots, r_{n+1})$$
$$+ (-1)^{n+1} f(r_1, \ldots, r_n) r_{n+1}\}.$$

1.19. Remark ([ML], X.3; [L₁], §1):

i. For each $n \in \mathbf{N}$ there exists a canonical isomorphism

$$H^n(R/k, M) \cong H^n(\mathrm{Hom}_{R^e}(\beta.(R/k), M)).$$

ii. There exist canonical isomorphisms

$$H^0(R/k, M) \cong \{m \in M : rm = mr \text{ for all } r \in R\}$$

and

$$H^1(R/k, M) \cong \mathrm{Der}_k(R, M)/\{\text{inner derivations}\}$$

where a k–derivation $d : R \to M$ is called an inner derivation if it is of the form $d(r) = rm - mr$ for some fixed $m \in M$.

iii. If M and N are two R–bimodules, then there exists a canonical cohomology product

$$H^p(R/k, M) \otimes H^q(R/k, N) \to H^{p+q}(R/k, M \otimes_{R^e} N).$$

Therefore if $\mu : M \otimes_{R^e} M \to M$ is a homomorphism of R^e–modules such that the corresponding multiplication $M \times M \to M$ is associative, then $H^\cdot(R/k, M) := \bigoplus_{n \in \mathbf{N}} H^n(R/k, M)$ carries the structure of an associative graded R-algebra. In terms of the defining complexes the cohomology product can be described as follows:

For cochains $f \in \mathrm{Mult}_k^p(R, M)$ and $g \in \mathrm{Mult}_k^q(R, N)$ define a $(p + q)$–cochain $f \otimes g \in \mathrm{Mult}_k^{p+q}(R, M \otimes_{R^e} N)$ by

$$f \otimes g(r_1, \ldots, r_{p+q}) = f(r_1, \ldots, r_p) \otimes g(r_{p+1}, \ldots, r_{p+q}).$$

This induces the desired product (c.f. [L₁], (1.8.1)).

Suppose now that $(R, \tau)/k$ is a topological algebra, and let M be an $R\hat{\otimes}_k R$-module which is complete in its $\tau \hat{\otimes} \tau$-adic topology. Then clearly M is also an R–bimodule. Let $\mathrm{Mult}_{k,\mathrm{cont}}^n(R, M) \subseteq \mathrm{Mult}_k^n(R, M)$ be the $\hat{R}^e$-submodule of continuous multilinear forms, i.e. the subset of all $f \in \mathrm{Mult}_k^n(R, M)$ such that for any open submodule $U \subset M$ there exists an open ideal $\mathfrak{I} \subset R$ such that f induces a well defined multilinear map $\overline{f} \in \mathrm{Mult}_k^n(R/\mathfrak{I}, M/U)$. We set $\mathrm{Mult}_{k,\mathrm{cont}}^0(R, M) := M$.

1.20. LEMMA. *Denoting by $\delta^{\cdot}$ the coboundary operator of $\underset{n\in\mathbb{N}}{\oplus} \operatorname{Mult}_k^n(R, M)$ as defined above, it holds:*

$$\delta^n(\operatorname{Mult}_{k,\text{cont}}^n(R, M)) \subseteq \operatorname{Mult}_{k,\text{cont}}^{n+1}(R, M)$$

Proof: Let $U \subseteq M$ be an open submodule, let $f \in \operatorname{Mult}_{k,\text{cont}}^n(R, M)$ and let $\mathfrak{I} \in \tau$ be an open ideal of R such that f induces $\bar{f} \in \operatorname{Mult}_k^n(R/\mathfrak{I}, M/U)$. Passing to a smaller open ideal $\mathfrak{I} \in \tau$ if necessary, we may assume that $(\mathfrak{I}\hat{\otimes}R + R\hat{\otimes}\mathfrak{I})M \subset U$, where by abuse of notation $\mathfrak{I}\hat{\otimes}R$ denotes the ideal generated by the image of $\mathfrak{I} \otimes_k R$ by the canonical map

$$\mathfrak{I} \otimes_k R \to R \otimes_k R \to R\hat{\otimes}_k R$$

and similar for $R\hat{\otimes}\mathfrak{I}$. But then $\delta^n f$ obviously induces $\overline{\delta^n f} \in \operatorname{Mult}_k^{n+1}(R/\mathfrak{I}, M/U)$, showing that $\delta^n f$ is continuous.

By the lemma we get a well defined subcomlex

$$(\underset{n\in\mathbb{N}}{\oplus} \operatorname{Mult}_{k,\text{cont}}^n(R, M), \delta^{\cdot}) \subset (\underset{n\in\mathbb{N}}{\oplus} \operatorname{Mult}_k^n(R, M), \delta^{\cdot})$$

1.21. DEFINITION. $H^n(R/k, \tau, M) := H^n((\underset{\ell\in\mathbb{N}}{\oplus} \operatorname{Mult}_{k,\text{cont}}^{\ell}(R, M), \delta^{\cdot}))$ *is called the nth topological Hochschild cohomology group of $(R, \tau)/k$ with coefficients in M.* $H^{\cdot}(R/k, \tau, M) := \underset{n\in\mathbb{N}}{\oplus} H^n(R/k, \tau, M)$.

Immediately from the definition we get the following properties of topological Hochschild cohomology groups:

1.22. Remark:

i. $H^n(R/k, \tau, \underline{\quad})$ defines a covariant functor on the category whose objects are the $\tau\hat{\otimes}\tau$–adic complete $\hat{R}^e$–modules, and whose morphisms are the continuous $\hat{R}^e$–homomorphisms.

ii. If τ is the discrete topology on R, then the above definition coincides with the classical definition of Hochschild cohomology. In this case we will write $H^{\cdot}(R/k, M)$ instead of $H^{\cdot}(R/k, \tau, M)$.

iii. We have canonical isomorphisms

$$H^0(R/k, \tau, M) \cong \{m \in M : rm = mr \text{ for all } r \in R\}$$

and

$$H^1(R/k, \tau, M) \cong \mathrm{Der}_{k,\mathrm{cont}}(R, M)/\{\text{inner derivations}\}.$$

In particular if τ is a topology such that for any open ideal $\mathfrak{I} \in \tau$ also $\mathfrak{I}^2 \subset R$ is open, then any k–derivation $d : R \to M$ is continuous, and therefore

$$H^1(R/k, \tau, M) \cong \mathrm{Der}_k(R, M)/\{\text{inner derivations}\}.$$

iv. The n–fold topological tensor product $T_k^n(R, \tau)$ of $(R, \tau)/k$ comes equipped with a canonical topology, and the universal properties of tensor product and completion imply the existence of canonical isomorphisms

$$\mathrm{Mult}_{k,\mathrm{cont}}^n(R, M) \cong \mathrm{Hom}_{k,\mathrm{cont}}(T_k^n(R, \tau), M).$$

If $\varphi : (R, \tau)/k \to (R', \tau')/k'$ is a morphism of topological algebras, then the induced homomorphisms $T^n(\varphi) : T_k^n(R, \tau) \to T_{k'}^n(R', \tau')$ are continuous. Therefore if M is an $R' \hat{\otimes} R'$–module which is complete both in its $\tau' \hat{\otimes} \tau'$–adic and in its $\tau \hat{\otimes} \tau$–adic topology then the $T^n(\varphi)$ induce canonical morphisms

$$H^n(R'/k', \tau', M) \to H^n(R/k, \tau, M).$$

In particular we always have canonical maps

$$H^n(R/k, \tau, M) \to H^n(R/k, M) \qquad (n \in \mathbb{N})$$

and canonical isomorphisms

$$H^n(\hat{R}/k, \hat{\tau}, M) \to H^n(R/k, \tau, M) \qquad (n \in \mathbb{N}).$$

As in the discrete case Hochschild cohomology groups carry a canonical R–module structure: The R–bimodule structure on M induces a right and a left R–module structure on $\mathrm{Mult}_{k,\mathrm{cont}}^n(R, M)$. Both of these structures pass to cohomology and induce a right and a left R–module structure on $H^n(R/k, \tau, M)$. These two structures coincide since for a given $r \in R$ the $R \hat{\otimes}_k R$–homomorphism

$$t_{n+1} : \mathrm{Mult}_{k,\mathrm{cont}}^{n+1}(R, M) \to \mathrm{Mult}_{k,\mathrm{cont}}^n(R, M)$$

given by the formula

$$t_{n+1}(f)(r_1, \ldots, r_n) = \sum_{i=1}^{n} (-1)^i f(r_1, \ldots, r_i, r, r_{i+1}, \ldots, r_n)$$

defines a homotopy between right multiplication by r and left multiplication by r.

If M and N are two $\tau \hat{\otimes} \tau$–adic complete $R \hat{\otimes}_k R$–modules, then we define for cochains $f \in \mathrm{Mult}^p_{k,\mathrm{cont}}(R, M)$ and $g \in \mathrm{Mult}^q_{k,\mathrm{cont}}(R, N)$ a cochain $f \hat{\otimes} g \in \mathrm{Mult}^{p+q}_{k,\mathrm{cont}}(R, M \otimes_{\hat{R}^e} N)$ by the formula

$$f \hat{\otimes} g(r_1, \ldots, r_{p+q}) = f(r_1, \ldots, r_p) \hat{\otimes} g(r_{p+1}, \ldots, r_{p+q}).$$

Calculations as in [$\mathbf{L_1}$], (1.8.1) show that this construction induces a cohomology product, and therefore we get

1.23. PROPOSITION. *If M and N are $\tau \hat{\otimes} \tau$–adic complete $R \hat{\otimes}_k R$–modules then there exists a canonical cohomology product*

$$H^p(R/k, \tau, M) \otimes_R H^q(R/k, \tau, N) \to H^{p+q}(R/k, \tau, M \otimes_{\hat{R}^e} N).$$

Therefore if $\mu : M \otimes_{\hat{R}^e} M \to M$ is a continuous homomorphism of $\hat{R}^e$–modules such that the corresponding multiplication $M \times M \to M$ is associative, then $H^{\cdot}(R/k, \tau, M)$ carries the structure of an associative graded R–algebra.

§2. Differential forms and Hochschild homology.

Hochschild homology and the algebra of differential forms of an algebra are closely connected. For an algebra R/k we denote by $\Omega^{\cdot}_{R/k}$ the algebra of Kähler differentials of R/k (universal differential algebra of R/k in the sense of [**KD**], (3.2), de Rham algebra). $\Omega^{\cdot}$ defines a covariant functor from A to ADG.

It is well known that there exists a natural transformation of functors

$$\theta^{\cdot} : \Omega^{\cdot} \to H.$$

([**L$_1$**], (1.10.2); see also [**HKR**] and [**R**]). Furthermore for every smooth algebra R/k of finite type

$$\theta^{\cdot}_{R/k} : \Omega^{\cdot}_{R/k} \to H.(R/k)$$

is an isomorphism ([**L$_1$**], (4.6.3); [**HKR**], p.395). More generally this is true, if R/k is a direct limit of smooth algebras which are essentially of finite type (see (2.13)).

In this section the above results will be generalized to topological algebras.

Let $(R, \tau)/(k, \tau_0)$ be a topological algebra and let (Ω, d) be a differential algebra of R/k. If $\mathfrak{J} \subseteq R$ is an ideal denote by $(\mathfrak{J}, d\mathfrak{J}) \subseteq \Omega$ the differentially closed two–sided ideal of Ω generated by $\mathfrak{J}$. For any two ideals $U, V \subseteq R$ it holds: $(U \cap V, d(U \cap V)) \subseteq (U, dU) \cap (V, dV)$. Therefore the ideals $(U, dU), U$ running through all open ideals of R, form a fundamental set of neighborhoods of 0 for a linear topology on Ω which will be denoted by $(\tau, d\tau)$.

For every ideal $U \subseteq R$ we have that $\Omega/(U, dU)$ together with the map induced by d is a differential algebra, and for any two ideals $U \subseteq V$ the canonical projection $\varphi_{V,U} : \Omega/(U, dU) \to \Omega/(V, dV)$ is a homomorphism of differential algebras. Therefore $\{(\Omega/(U, dU), \varphi_{V,U}) : U \subseteq R$ is an open ideal$\}$ is a projective system of differential algebras. Since the category of differential algebras is complete ([**KD**], (2.11)) there exists the projective limit of this system in the category of differential algebras. It will be denoted by $\widehat{\Omega_\tau}$. The explicit construction of products and equalizers in [**KD**], proof of (2.11), and the description of limits in [**Sch**], thm. (7.4.2), show that $\widehat{\Omega_\tau}$ is a differential algebra of $\hat{R}/k$. Furthermore there exists a canonical morphism $i : \Omega \to \widehat{\Omega_\tau}$ of differential algebras.

2.1. DEFINITION. $\widehat{\Omega_\tau}$ is called *completion of the differential algebra Ω with respect to the topology* $(\tau, d\tau)$.

A differential algebra Ω *of* R/k *is called* $(\tau, d\tau)$*-complete, if* $i : \Omega \to \widehat{\Omega_\tau}$ *is an isomorphism. In particular* (R, τ) *is complete in this case.*

Ω is also a topological ring with respect to the topology $(\tau, d\tau)$, and d is a continuous homomorphism of groups with respect to this topology. Let $\hat{\Omega}$ be the completion of Ω as a ring (resp. group), and let $\hat{d}$ be the continuous continuation of the map induced by d on $im(\Omega \to \hat{\Omega})$ to $\hat{\Omega}$. Clearly $\hat{\Omega}$ is an $\hat{R}$–algebra, however in general it is not a DG–algebra. We have:

2.2. LEMMA. *There exists an injective homomorphism*

$$\varphi : \widehat{\Omega_\tau} \to \hat{\Omega}$$

of $\hat{R}$*–algebras satisfying* $\varphi \circ d = \hat{d} \circ \varphi$.

This morphism maps $\widehat{\Omega_\tau}$ *isomorphically onto the subalgebra* $\hat{R}\left[d\hat{R}\right]$ *of* $\hat{\Omega}$.

φ *is an isomorphism if and only if* $\hat{\Omega}$ *is a differential algebra of* $\hat{R}/k$.

Proof: Using the explicit construction of products and equalizers in the category of differential algebras as subobjects of products and equalizers in the category of rings ([**KD**], proof of (2.11)) and the description of limits in [**Sch**], thm. (7.4.2), one sees easily that $\widehat{\Omega_\tau}$ is canonically isomorphic to the subalgebra $\hat{R}\left[d\hat{R}\right]$ of $\hat{\Omega}$, and that d is the map induced by $\hat{d}$, proving the first part of (2.2).

The second part follows immediately from the first one.

Remark: The lemma shows in particular that completions of differential algebras as considered in [**KD**], §12 are a special case of completions of differential algebras as defined in (2.1).

Notation: If $\Omega = \Omega_{R/k}$ is the universal differential algebra of R/k, then we denote $\widehat{\Omega_\tau}$ by $\Omega_{(R/k,\tau)}$ and call it the *universally complete differential algebra of* $(R, \tau)/k$.

2.3. *Remark*: There exists a canonical isomorphism of differential algebras

$$\Omega_{(R/k,\tau)} \to \Omega_{(\hat{R}/k,\hat{\tau})}.$$

Proof: If $\mathfrak{A}$ is a fundamental set of neighborhoods of 0 in R, consisting of ideals of R, then $\widehat{\mathfrak{A}} := \left\{ \overline{U} \subseteq \hat{R} : U \in \mathfrak{A} \right\}$ is a fundamental set of neighborhoods of 0 in $\hat{R}$. Here $\overline{U}$ denotes the topological closure of the image of U in $\hat{R}$. Since we have canonical isomorphisms

$$\Omega^{\cdot}_{R/k}/(U,dU) = \Omega^{\cdot}_{R/U/k} = \Omega^{\cdot}_{\hat{R}/\overline{U}/k} = \Omega^{\cdot}_{\hat{R}/k}/\left(\overline{U}, d\overline{U} \right) \ (U \in \mathfrak{A})$$

the claim follows.

2.4. PROPOSITION.

i. Let Ω be a differential algebra of $(R, \tau)/k$, and let Ω' be a differential algebra of $(R', \tau')/k'$. Suppose that $\varphi : \Omega \to \Omega'$ is a continuous homomorphism of differential algebras and that Ω' is $(\tau, d\tau)$-complete. Then there exists a unique homomorphism

$$\hat{\varphi} : \widehat{\Omega_{\tau}} \to \Omega'$$

of differential algebras satisfying $\hat{\varphi} \circ i = \varphi$.

In particular every $(\hat{\tau}, d\hat{\tau})$-complete differential algebra of $\hat{R}/k$ is a homomorphic image of $\Omega^{\cdot}_{(R/k, \tau)}$.

ii. $\Omega^{\cdot}$ defines a covariant functor

$$\Omega^{\cdot} : \ Topal \to ADG$$

whose restriction to the full subcategory A of commutative algebras is the classical functor $\Omega^{\cdot}$.

Proof: i. Denote by $\hat{\Omega}$ resp. $\widehat{\Omega'}$ the $(\tau, d\tau)$-completion resp. the $(\tau', d\tau')$-completion of the rings Ω resp. Ω'. Being a continuous homomorphism of rings φ induces a map

$$\tilde{\varphi} : \hat{\Omega} \to \widehat{\Omega'}$$

making the following diagram commutative:

$$
\begin{array}{ccc}
\hat{\Omega} & \xrightarrow{\ \tilde{\varphi}\ } & \widehat{\Omega'} \\[4pt]
\Big\uparrow{\scriptstyle can.} & & \Big\uparrow{\scriptstyle can.} \\[4pt]
\Omega & \xrightarrow{\ \varphi\ } & \Omega'.
\end{array}
$$

Identifying $\widehat{\Omega}_\tau$ with the subalgebra $\hat{R}\left[\hat{d}\hat{R}\right]$ of $\hat{\Omega}$ according to (2.2) we get the desired morphism $\hat{\varphi}$ by restricting $\tilde{\varphi}$, since $\tilde{\varphi}$ commutes with $\hat{d}$, and since Ω' is isomorphic to the subalgebra $R'\left[\hat{d}R'\right] = \widehat{R'}\left[\widehat{dR'}\right]$ of $\widehat{\Omega'}$. Clearly this implies that every $(\hat{\tau}, d\hat{\tau})$–complete differential algebra of $\hat{R}/k$ is a homomorphic image of $\Omega_{(\hat{R}/k,\hat{\tau})}$, hence by (2.3) of $\Omega_{(R/k,\tau)}$.

ii. is a formal consequence of i.

Given suitable assumptions about $(R,\tau)/k$ we can relate $\Omega_{(R/k,\tau)}$ to well–known objects.

2.5. Remark: Let R be noetherian and let τ be the $\mathfrak{J}$–adic topology on R for some ideal $\mathfrak{J} \subseteq R$.

i. Suppose that the universally finite differential algebra $\widetilde{\Omega}_{\hat{R}/k}$ of $\hat{R}/k$ in the sense of [KD], (11.3) exists. Then $\widetilde{\Omega}_{\hat{R}/k}$ is a homomorphic image of $\Omega_{(R/k,\tau)}$.

ii. $\Omega_{(R/k,\tau)}$ is the universally finite differential algebra of $\hat{R}/k$ if and only if $\Omega^1_{(R/k,\tau)}$ is finite.

Proof: For an arbitrary differential algebra Ω of $\hat{R}/k$ and any ideal $\mathfrak{A} \subseteq \hat{R}$ it holds: $d(\mathfrak{A}^n\Omega) \subseteq \mathfrak{A}^{n-1}\Omega$. In particular we have:

$$\mathfrak{J}^n\Omega \subseteq \left(\mathfrak{J}^n\hat{R}, d\left(\mathfrak{J}^n\hat{R}\right)\right) \subseteq \mathfrak{J}^{n-1}\Omega \subseteq \left(\mathfrak{J}^{n-1}\hat{R}, d\left(\mathfrak{J}^{n-1}\hat{R}\right)\right)$$

for every $n \geq 1$, implying that the topology $(\hat{\tau}, d\hat{\tau})$ on Ω coincides with the $\mathfrak{J}\hat{R}$–adic topology on Ω. Since every finite $\hat{R}$–module is complete in its $\mathfrak{J}\hat{R}$–adic topology the statements of (2.5) follow immediately from (2.4).

Working with differential modules of topological algebras it sometimes makes sense to look at the completion $\hat{\Omega}^1_{R/k}$ of the R–module $\Omega^1_{R/k}$ with respect to the topology $(\tau, d\tau) \cap \Omega^1_{R/k}$ (c.f. [EGA $\mathbf{O_{IV}}$], (20.7.14)). By [$\mathbf{B_3}$], II, §6.1, Prop. 1 there exists a canonical injection

$$\hat{\Omega}^1_{R/k} \to \hat{\Omega}'_{R/k}$$

where $\hat{\Omega}'_{R/k}$ is the $(\tau, d\tau)$–completion of the ring $\Omega'_{R/k}$. We have:

2.6. PROPOSITION.

i. After identifying $\Omega_{(R/k,\tau)}$ and $\hat{\Omega}^1_{R/k}$ with their images in $\hat{\Omega}'_{R/k}$ by the canonical maps it holds: $\Omega^1_{(R/k,\tau)} \subseteq \hat{\Omega}^1_{R/k}$.

Equality holds if and only if $\hat{\Omega}^1_{R/k}$ is generated by the elements of the form $dr, r \in \hat{R}$ as an $\hat{R}$–module.

ii. Let R be noetherian and τ be the $\mathfrak{I}$–adic topology on R for some ideal $\mathfrak{I} \subseteq R$. If $\Omega'_{R/\mathfrak{I}/k}$ is a finite $R/\mathfrak{I}$–module then the universally finite differential algebra of $\hat{R}/k$ exists and is canonically isomorphic to $\Omega'_{(R/k,\tau)}$.

Proof: i. is an easy consequence of the various definitions.

ii. By [**EGA 0_{IV}**], (20.7.15) $\hat{\Omega}^1_{R/k}$ is a finite $\hat{R}$–module, and so is $\Omega^1_{(R/k,\tau)}$ by i,. The statement follows now from (2.5) ii,.

Under suitable assumptions $\Omega^1_{(R/k,\tau)}$ can be described by $\ker\left(R\widehat{\otimes}_k R \overset{\mu}{\to} \hat{R}\right)$ in analogy to the description of $\Omega^1_{R/k}$ by $\ker\left(R \otimes_k R \overset{\mu}{\to} R\right)$.

2.7. PROPOSITION. *Let $(R,\tau)/k$ be a topological algebra satisfying*

i. R is noetherian and τ is the $\mathfrak{K}$–adic topology on R for some ideal $\mathfrak{K} \subseteq R$.

*ii. The complete tensor product $R\widehat{\otimes}_k R$ is noetherian. (If k is noetherian, and if $R/\mathfrak{K}$ is essentially of finite type over k, then i. implies ii. by [**N**], thm. (17.3) and [**AM**], (10.25).)*

If $\mathfrak{J} := \ker\left(R\widehat{\otimes}_k R \to R\right)$ then there exists a canonical isomorphism

$$\Omega^1_{(R/k,\tau)} \cong \mathfrak{J}/\mathfrak{J}^2.$$

Proof: By (2.3) we may assume: R is $\mathfrak{K}$–adically complete.
Assumption ii. implies that $\Omega'_{R/\mathfrak{K}/k}$ is a finite $R/\mathfrak{K}$–module so that it suffices to show that $\mathfrak{J}/\mathfrak{J}^2$ has the universal property of the universally finite differential module of R/k.
In the following we denote by $x\hat{\otimes}y$ the image of $x \otimes y$ by the canonical map

$$i : R \otimes_k R \to R\widehat{\otimes}_k R.$$

By (1.5), the topology $\tau\hat{\otimes}\tau$ on $R\widehat{\otimes}_k R$ is the $\mathfrak{K}_{(2)} \cdot R\widehat{\otimes}_k R$–adic topology, $\mathfrak{K}_{(2)}$ being the ideal defined in (1.8). Since $R\widehat{\otimes}_k R$ is noetherian, the subspace topology on $\mathfrak{J} \subseteq R\widehat{\otimes}_k R$ is the $\mathfrak{K}_{(2)} \cdot R\widehat{\otimes}_k R$–adic topology on $\mathfrak{J}$ by the Artin–Rees–Lemma, and therefore the topology induced on the quotient $\mathfrak{J}/\mathfrak{J}^2$ is the $\mathfrak{K}$–adic topology.

1) Every $x \in \mathfrak{J}$ has a representation $x = \sum\limits_{n\in\mathbb{N}} x_n$ as a convergent series in the $\mathfrak{K}_{(2)} R\widehat{\otimes}_k R$–adic topology with $x_n = \sum\limits_{\text{finite}} a_{n,i}\hat{\otimes}1 \cdot \left(b_{n,i}\hat{\otimes}1 - 1\hat{\otimes}b_{n,i}\right) (a_{n,i}, b_{n,i} \in R; n \in \mathbb{N})$, because:

There exists a sequence $(x_n) \in R \otimes_k R$ such that $x - i(x_n) \in (\mathfrak{K}^n \otimes R + R \otimes \mathfrak{K}^n) \cdot R \widehat{\otimes}_k R$.
Consequently $\mu(x_n) = \mu(i(x_n) - x) \in \mathfrak{K}^n$, hence $\mu(x_n) \hat{\otimes} 1 \in (\mathfrak{K}^n \otimes R + R \otimes \mathfrak{K}^n) \cdot R \widehat{\otimes}_k R$.
So $i(x_n) - \mu(x_n) \hat{\otimes} 1$ also converges to x, and $i(x_n) - \mu(x_n) \hat{\otimes} 1 \in i(\mathfrak{A})$ for all $n \in \mathsf{N}$ with
$\mathfrak{A} := \ker\left(R \otimes_k R \xrightarrow{\mu} R\right)$.
Therefore we may assume: $x_n \in \mathfrak{A}$ for all $n \in \mathsf{N}$.
Then $x = i(x_0) + \sum_{n \geq 1} i(x_n - x_{n-1})$ as convergent series, and $x_0, x_n - x_{n-1} \in \mathfrak{A}$ for all $n \in \mathsf{N}$.
It is well known:

$$x_0 = \sum_{\text{finite}} a_{0,i} \otimes 1 \cdot (b_{0,i} \otimes 1 - 1 \otimes b_{0,i})$$

$$x_n - x_{n-1} = \sum_{\text{finite}} a_{n,i} \otimes 1 \cdot (b_{n,i} \otimes 1 - 1 \otimes b_{n,i}) \quad (n \geq 1)$$

with certain $a_{n,i}, b_{n,i} \in R$ for all $n \in \mathsf{N}$.

2) As an R–module $\mathfrak{J}/\mathfrak{J}^2$ is generated by elements of the form $dr := r \hat{\otimes} 1 - 1 \hat{\otimes} r + \mathfrak{J}^2$
$r \in R$, because:

By 1) the elements $dr, r \in R$ generate a submodule M of $\mathfrak{J}/\mathfrak{J}^2$ which is dense in the $\mathfrak{K}$–adic topology. As a submodule of the finite R–module $\mathfrak{J}/\mathfrak{J}^2$, M is finite as an R–module, hence complete in its $\mathfrak{K}$–adic topology by the Artin–Rees–Lemma, and we conclude $M = \mathfrak{J}/\mathfrak{J}^2$.

3) $d : R \to \mathfrak{J}/\mathfrak{J}^2 \left(r \mapsto r \hat{\otimes} 1 + 1 \hat{\otimes} r + \mathfrak{J}^2\right)$ is a well–defined k–derivation.

4) Let N be a finite R–module and let $\delta \in \mathrm{Der}_k(R, N)$. Then there exists a unique R–homomorphism

$$h : \mathfrak{J}/\mathfrak{J}^2 \to N$$

making the following diagram commute

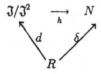

because:

$R \ltimes N$, the ring of dual numbers, is $\mathfrak{K}$–adically complete since it is finite as an R–module, and

$$\psi_1 : R \to R \ltimes N (r \mapsto (r, 0))$$

$$\psi_2 : R \to R \ltimes N (r \mapsto (r, \delta(r)))$$

are continuous homomorphisms of k–algebras. By the universal property of complete tensor products (1.6) there exists a unique continuous homomorphism of topological k–algebras

$$\psi : R\widehat{\otimes}_k R \to R \ltimes N$$

satisfying $\psi\left(r\hat{\otimes}s\right) = (rs, r\delta(s))$.

Clearly $\psi(\mathfrak{J}) \subseteq N$, hence $\psi\left(\mathfrak{J}^2\right) \subseteq N^2 = (0)$. Therefore ψ induces a homomorphism

$$h : \mathfrak{J}/\mathfrak{J}^2 \to N$$

which is easily seen to be R–linear, and which satisfies: $h(dr) = h\left(1\hat{\otimes}r - r\hat{\otimes}1 + \mathfrak{J}^2\right) = \delta(r) - r\delta(1) = \delta(r)$ for all $r \in R$. 2) implies the uniqueness of h.

1)–4) show that $\left(\mathfrak{J}/\mathfrak{J}^2, d\right)$ has the universal property of the universally finite differential module of R/k, completing the proof of the proposition.

Next the relations between the two functors

$$\Omega^{\cdot}, H. : \text{Topal} \to ADG$$

"differential algebras" and "Hochschild homology" will be examined.

Let $(R, \tau)/(k, \tau_0)$ be a topological algebra and denote by d the differentiation on $H.(R/k, \tau)$. The subalgebra $\hat{R}\left[d\hat{R}\right]$ of $H.(R/k, \tau) = H.\left(\hat{R}/k, \tau\right)$ is a differential algebra of $\hat{R}/k$ since $dr \cdot dr = 0$ for every $r \in R$ as was shown in the proof of (1.11). Let Topal$^\sim$ be the full subcategory of Topal whose objects satisfy: $\hat{R}\left[d\hat{R}\right]$ is $(\tau, d\tau)$–complete in the sense of definition 2.1.

2.8. *Remark:* Let R be noetherian and let τ be the $\mathfrak{J}$–adic topology for an ideal $\mathfrak{J} \subseteq R$. If $H_1(R/k, \tau)$ is a finite $\hat{R}$–module, then $(R, \tau)/(k, \tau_0)$ is in Topal$^\sim$.

In this case the topology $(\hat{\tau}, d\hat{\tau})$ on a differential algebra Ω of $\hat{R}$ coincides with the $\mathfrak{J}\hat{R}$–adic topology on Ω. The assumptions imply that $\hat{R}\left[d\hat{R}\right] \subseteq H.(R/k, \tau)$ is $\mathfrak{J}\hat{R}$–adically complete as an $\hat{R}$–module. By (2.2) this implies that the differential algebra $\hat{R}\left[d\hat{R}\right]$ is $(\hat{\tau}, d\hat{\tau})$–complete.

Topal$^\sim$ contains the full subcategory A of commutative algebras. In addition it contains the following important objects:

2.9. PROPOSITION. Let $(R, \tau)/(k, \tau_0)$ be a topological algebra having the following properties:

i) R is noetherian and τ is the $\mathfrak{I}$–adic topology on R for some ideal $\mathfrak{I} \subseteq R$.

ii) $T_k^n(R, \tau)$ is noetherian for $n \leq 4$.

Furthermore let $\mathfrak{J} := \ker\left(R \hat{\otimes}_k R \to \hat{R}\right)$. Then there exists an isomorphism of $\hat{R}$–modules

$$H_1(R/k, \tau) \xrightarrow{\sim} \mathfrak{J}/\mathfrak{J}^2$$

and $(R, \tau)/(k, \tau_0)$ is an object of $\text{Topal}^{\sim}$.

Proof: We may assume that R is $\mathfrak{I}$–adically complete.

Using (2.8) it suffices to show that $H_1(R/k, \tau) \cong \mathfrak{J}/\mathfrak{J}^2$ since $\mathfrak{J}/\mathfrak{J}^2$ is a finite R-module by assumption ii). Let $(\beta.(R/k, \tau), \partial.)$ be the topological bar resolution of $(R, \tau)/(k, \tau_0)$. Then $\mathfrak{J} = im(\partial_1)$, and by (1.5)iv) and vi) there exist canonical isomorphisms of $\widehat{R^e}$–modules

$$R \otimes_{\widehat{R^e}} \beta_n(R/k, \tau) \cong \gamma_n(R/k, \tau) \qquad (n \leq 2)$$

By this isomorphism $1 \otimes \partial_n$ corresponds to $d_n(n \leq 2)$. Therefore we get a commutative diagram with the middle column being exact:

$$
\begin{array}{ccc}
R \otimes_{\widehat{R^e}} \beta_2(R/k, \tau) & = & \gamma_2(R/k, \tau) \\
\downarrow{\scriptstyle 1 \otimes \partial_2} & & \downarrow{\scriptstyle d_2} \\
R\widehat{\otimes}_k R = R \otimes_{\widehat{R^e}} \beta_1(R/k, \tau) & = & \gamma_1(R/k, \tau) \\
\downarrow \qquad \downarrow{\scriptstyle can} \;{\scriptstyle 1 \otimes \partial_1} & & \downarrow{\scriptstyle d_1} \\
\mathfrak{J}/\mathfrak{J}^2 = R \otimes_{\widehat{R^e}} \mathfrak{J} & \longrightarrow & \gamma_0(R/k, \tau) \\
\downarrow & & \\
0 & &
\end{array}
$$

Since $d_1 = 0$ this implies:

$$H_1(R/k, \tau) = \ker(d_1)/im(d_2) = \gamma_1(R/k, \tau)/im(1 \otimes \partial_2) = \mathfrak{J}/\mathfrak{J}^2.$$

2.10. THEOREM. *There exists a natural transformation of functors*

$$\theta^{\cdot} : \Omega^{\cdot}|_{\text{Topal}^{\sim}} \rightarrow H.|_{\text{Topal}^{\sim}}$$

whose restriction to the category of commutative algebras is the transformation of ([$\mathbf{L_1}$], *(1.10.2)).*

Proof: Let $(R, \tau)/(k, \tau_0)$ be in Topal$^{\sim}$. The subalgebra $A := \hat{R}\left[d\hat{R}\right]$ of $H.(R/k, \tau)$ is $(\hat{\tau}, d\hat{\tau})$–complete as differential algebra of $\hat{R}/k$ by assumption. Therefore there exists by (2.4) a unique epimorphism of differential algebras of $\hat{R}/k$

$$\Omega^{\cdot}_{(R/k,\tau)} \rightarrow A.$$

Composed with the inclusion $A \subseteq H.(R/k, \tau)$ this gives an $\hat{R}$–linear homomorphism of anti-commutative DG–algebras

$$\theta^{\cdot}_{(R/k,\tau)} : \Omega^{\cdot}_{(R/k,\tau)} \rightarrow H.(R/k, \tau).$$

Clearly this construction is functorial in $(R, \tau)/(k, \tau_0)$. If τ is the discrete topology on R, then $\theta^{\cdot}_{(R/k,\tau)}$ is given by

$$\theta^{\cdot}_{(R/k,\tau)} (r_0 dr_1 \cdot \ldots \cdot dr_n)$$
$$= \theta^{\cdot}_{(R/k,\tau)} (r_0) \cdot \theta^{\cdot}_{(R/k,\tau)} (dr_1) \cdot \ldots \cdot \theta^{\cdot}_{(R/k,\tau)} (dr_n)$$
$$= \text{homology class of} \sum_{\sigma \in S_n} \text{sign}\,(\sigma) r_0 \otimes r_{\sigma(1)} \otimes \cdots \otimes r_{\sigma(n)}$$

for $r_0, r_1, \ldots, r_n \in R, n \in \mathbf{N}$. Hence it coincides with the map constructed in ($\mathbf{L_1}$, (1.10.2)). In this case it will be denoted by $\theta^{\cdot}_{R/k}$.

In some cases it can be shown that $\theta^{\cdot}$ is an isomorphism.

2.11. PROPOSITION. *Let k be noetherian and let R/k be essentially of finite type and smooth (in the sense of [KD], §8*). Then*

$$\theta^{\cdot}_{R/k} : \Omega^{\cdot}_{R/k} \rightarrow H.(R/k)$$

---

*E. Kunz has asked me to mention the following corrections to [KD]: The equivalences of the statements a), b) and c) of theorem 8.1 require S/R to be of finite type. The corollaries remain correct for algebras which are essentially of finite type, however their proofs have to be changed slightly. In Proposition 8.7 and theorem 10.12 of [KD] one has to assume, too, that S/R is of finite type.

is an isomorphism of DG–algebras.

Proof: Since R/k is flat we may assume R/k is a local algebra by the functoriality of θ', by (1.15) and by [**KD**], (4.21). Then $R = S_{\mathfrak{P}}$ for some algebra S/k of finite type and for some $\mathfrak{P} \in \mathrm{Spec}(S)$ such that S/k is smooth at $\mathfrak{P}$. The smooth locus of S/k is open ([**KD**], (8.2)) and contains $\mathfrak{P}$. Therefore there exists an $f \in S$ such that $\mathfrak{P} \in D(f)$ and such that S_f/k is smooth. Since S_f/k is also of finite type we may assume: S/k is smooth.

Then $\theta'_{S/k} : \Omega'_{S/k} \to H.(S/k)$ is an isomorphism by ([**L₁**], (4.6.3)). Since $\theta'_{R/k}$ is the localization of $\theta'_{S/k}$ at $\mathfrak{P}$ ((2.10), (1.15) and [**KD**], (4.21)) this implies that

$$\theta'_{R/k} : \Omega'_{R/k} \to H.(R/k)$$

is an isomorphism of DG–algebras.

Let k be an arbitrary commutative ring and let Ω be a differential algebra of k. Furthermore let $(R_\lambda)_{\lambda \in \Lambda}$ be a direct system of k–algebras, and let $R := \varinjlim_{\lambda \in \Lambda} R_\lambda$ be its direct limit in the category of k–algebras. By the functorial property of universal extensions ([**KD**], §3) we get a corresponding system $(\Omega_{R_\lambda})_{\lambda \in \Lambda}$ of differential algebras. By [**KD**], (2.11) its direct limit $\varinjlim_{\lambda \in \Lambda} \Omega_{R_\lambda}$ exists in the category $\mathcal{D}_{\mathbf{Z}}$ of differential algebras over $\mathbf{Z}$, and the proof of [**KD**], (2.11) shows that $\varinjlim_{\lambda \in \Lambda} \Omega_{R_\lambda}$ is a differential algebra of R.

Let $i_\lambda : R_\lambda \to R$ and $j_\lambda : \Omega_{R_\lambda} \to \varinjlim_{\lambda \in \Lambda} \Omega_{R_\lambda}$ be the canonical morphisms into the direct limit. Then the composition $\varphi_R : \Omega \to \Omega_{R_\lambda} \to \varinjlim_{\lambda \in \Lambda} \Omega_{R_\lambda}$ is independent of λ since it coincides in degree 0 with the structural homomorphism $\rho : k \to \varinjlim_{\lambda \in \Lambda} R_\lambda = R$. In particular φ_R is a ρ–homomorphism.

2.12. LEMMA.

i) The abelian group of $\varinjlim_{\lambda \in \Lambda} \Omega_{R_\lambda}$ is the direct limit of $(\Omega_{R_\lambda})_{\lambda \in \Lambda}$ in the category of abelian groups.

ii) $\left(\varinjlim_{\lambda \in \Lambda} \Omega_{R_\lambda}, \varphi_R \right)$ is the universal R–extension of Ω in the sense of [**KD**], (3.11).

Proof: i) is an immediate consequence of the proof of [**KD**], (2.11).

ii) Let Ω^* be a differential algebra of R and let $\psi : \Omega \to \Omega^*$ be a ρ–homomorphism. By the universal property of Ω_{R_λ} there exists an i_λ–homomorphism $\psi_\lambda : \Omega_{R_\lambda} \to \Omega^*$. If $\mu : \Omega_{R_\lambda} \to \Omega_{R_\mu}$ is a homomorphism in the direct system $(\Omega_{R_\lambda})_{\lambda \in \Lambda}$, then the diagram

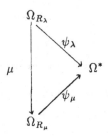

commutes, since it commutes in degree 0.

Therefore there exists an R–homomorphism

$$h : \varinjlim_{\lambda \in \Lambda} \Omega_{R_\lambda} \to \Omega^*$$

by the universal property of direct limits. Clearly $\psi = h \circ \varphi_R$, and it follows: $\varinjlim_{\lambda \in \Lambda} \Omega_{R_\lambda}$ is the universal R–extension of Ω.

2.13. COROLLARY. *Let k be noetherian and let $R := \varinjlim_{\lambda \in \Lambda} R_\lambda$ with $(R_\lambda/k)_{\lambda \in \Lambda}$ being a direct system of smooth algebras which are essentially of finite type over k. Then*

$$\theta_{R/k} : \Omega_{R/k}^{\cdot} \to H_\cdot(R/k)$$

is an isomorphism of DG–algebras.

In particular this holds, if R is a finite direct product of fields which are separable over k.

Proof: By (2.11) $\theta_{R_\lambda/k} : \Omega_{R_\lambda/k}^{\cdot} \to H_\cdot(R_\lambda/k)$ is an isomorphism of DG–algebras for every $\lambda \in \Lambda$. The $\theta_{R_\lambda/k}, \lambda \in \Lambda$ are a direct system of morphisms by (2.10). Passing to the direct limit yields the claim (using (2.10), (1.17) and (2.12)).

For proving the second part of (2.13) we may assume by (1.18), [**KD**], (4.7) and (2.10) that R/k is a separable field extension. Then $R = \varinjlim Z$, where Z ranges over all intermediate fields of R/k which are algebraic function fields over k. These Z are separable over k ([**KD**], (5.14)), hence in particular smooth.

2.14. COROLLARY. *Let k be noetherian and let R/k be essentially of finite type and smooth. Furthermore let τ be the $\mathfrak{I}$–adic topology on R for some ideal $\mathfrak{I} \subseteq R$. Then*

$$\theta^{\cdot}_{(\hat{R}/k, \hat{\tau})} : \Omega^{\cdot}_{(\hat{R}/k, \hat{\tau})} \to H.\left(\hat{R}/k, \hat{\tau}\right)$$

is an isomorphism of DG–algebras.

Proof: (2.7) and [**KD**],(12.5) imply:

$$\Omega^{\cdot}_{(R/k,\tau)} \cong \widetilde{\Omega}^{\cdot}_{\hat{R}/k} \cong \Omega^{\cdot}_{R/k} \otimes_R \hat{R}.$$

Since both R/k and $\widehat{R^e}/R^e$ are flat there exist canonical isomorphisms:

$$
\begin{array}{lll}
H.(R/k, \tau) & & \\
\cong & Tor^{\widehat{R^e}}.\left(\hat{R}, \hat{R}\right) & ((1.13)) \\
\cong & Tor^{\widehat{R^e}}.\left(\widehat{R^e} \otimes_{R^e} R, \widehat{R^e} \otimes_{R^e} R\right) & \\
\cong & Tor^{R^e}.(R, R) \otimes_{R^e} \widehat{R^e} & ([\mathbf{Ma}], 3.E) \\
\cong & H.(R/k) \otimes_R \hat{R}. &
\end{array}
$$

Since $\theta^{\cdot}_{R/k} : \Omega^{\cdot}_{R/k} \to H.(R/k)$ is an isomorphism by (2.11), the claim follows by the functoriality of $\theta^{\cdot}$.

2.15. EXAMPLE. *Let k be noetherian.*

a) *Let $R = k[[X_1, \ldots, X_n]]$ be the ring of formal power series in n variables over k, and let τ be the $(X_1, \ldots, X_n)$–adic topology on R. Then*

$$\theta^{\cdot}_{(R/k,\tau)} : \Omega^{\cdot}_{(R/k,\tau)} \to H.(R/k, \tau)$$

is an isomorphism of DG–algebras.

b) *Let $\mathfrak{I} \subseteq k$ be an ideal and let τ_0 be the $\mathfrak{I}$–adic topology on k. If $R = k\{X_1, \ldots, X_n\}$ is the ring of restricted power series with respect to τ_0 ([**EGA O$_I$**] (7.5); [**B$_1$**], III, §4.2), and if τ is the $\mathfrak{I} \cdot (X_1, \ldots, X_n)$–adic topology on R, then*

$$\theta^{\cdot}_{(R/k,\tau)} : \Omega^{\cdot}_{(R/k,\tau)} \to H.(R/k, \tau)$$

is an isomorphism of DG–algebras ([Sal], Prop. 1).

In Topal$^\sim$ let Topal$^\approx$ be the full subcategory whose objects $(R,\tau)/(k,\tau_0)$ have the following properties:

The homogeneous components $\Omega^n_{(R/k,\tau)}$ of $\Omega^{\cdot}_{(R/k,\tau)}$ are the R–module completions of $\Omega^n_{R/k}$ with respect to the topology $(\tau, d\tau) \cap \Omega^n_{R/k}$.

2.16. Remark:

i) Let $(R,\tau)/(k,\tau_0)$ be an object of Topal$^\sim$, and let $\mathfrak{A}$ be a fundamental set of neighborhoods of 0, consisting of ideals of R. Then $(R,\tau)/(k,\tau_0)$ is in Topal$^\approx$ if and only if

$$\Omega^n_{(R/k,\tau)} = \varprojlim_{U \in \mathfrak{A}} \Omega^n_{R/U/k} \text{ for all } n \in \mathsf{N}.$$

ii) Topal$^\approx$ contains the full subcategory of commutative algebras, and also those objects of Topal which satisfy the conditions of (2.7).

For objects of Topal$^\approx$ more information about the relations between Hochschild homology and differential forms can be obtained. The following proposition was first proved by Lipman ([L$_1$], (4.6.4)) in case τ is the discrete topology.

2.17. PROPOSITION. *For every* $(R,\tau)/(k,\tau_0)$ *in Topal$^\approx$ there exists a functorial homomorphism of graded $\hat{R}$–modules*

$$\overline{\delta}_{(R/k,\tau)} : H.(R/k,\tau) \to \Omega^{\cdot}_{(R/k,\tau)}$$

having the following properties:

i) $\overline{\delta}^n_{(R/k,\tau)} \circ \theta^n_{(R/k,\tau)} = n! \cdot id_{\Omega^n_{(R/k,\tau)}}$ *for all* $n \in \mathsf{N}$.

ii) $(n+1)d^n_{R/k} \circ \overline{\delta}^n_{(R/k,\tau)} = \overline{\delta}^{n+1}_{(R/k,\tau)} \circ \Delta^n_{R/k}$ *for all* $n \in \mathsf{N}$, *where* $d_{R/k}$ *resp.* $\Delta_{R/k}$ *denotes the differentiation on* $\Omega^{\cdot}_{(R/k,\tau)}$ *resp.* $H.(R/k,\tau)$.

iii) *It holds for* $\omega \in \Omega^m_{(R/k,\tau)}$ *and* $x \in H_n(R/k,\tau)$:

$$\overline{\delta}_{(R/k,\tau)}\left(\theta^{\cdot}_{(R/k,\tau)}(\omega) \cdot x\right) = \frac{(n+m)!}{n!} \cdot \omega \cdot \overline{\delta}_{(R/k,\tau)}(x).$$

Proof: Let $\mathfrak{A}$ be a fundamental set of neighborhoods of $0 \in R$, consisting of ideals of R.

For every $n \in \mathbb{N}$ there exists a projective system of morphisms of R/U–modules

$$f_U^n : \gamma_n(R/U/k) \to \Omega_{R/U/k}^n \ (U \in \mathfrak{A})$$

defined by

$$f_U^n (r_0 \otimes r_1 \otimes \cdots \otimes r_n) = r_0 dr_1 \cdot \ldots \cdot dr_n \ (r_0, r_1, \ldots, r_n \ \in R/U)$$

satisfying $f_U^n \circ d = 0$.

Since $(R/k, \tau)$ is in Topal$^{\approx}$, we get a well–defined homomorphism

$$f^n := \varprojlim_{U \in \mathfrak{A}} f_U^n : \gamma_n(R/k, \tau) \to \Omega_{(R/k,\tau)}^n$$

of $\hat{R}$–modules for every $n \in \mathbb{N}$ ((2.16)i), and it holds: $f^n \circ d = 0$ for every $n \in \mathbb{N}$.

Therefore $f^{\cdot} = \bigoplus_{n \in \mathbb{N}} f^n$ induces a well–defined mapping

$$\overline{\delta}_{(R/k,\tau)} : H_{\cdot}(R/k, \tau) \to \Omega_{(R/k,\tau)}^{\cdot}$$

of graded $\hat{R}$–modules. Clearly this construction is functorial in $(R, \tau)/(k, \tau_0)$.

i) If τ is the discrete topology on R, then $\overline{\delta}_{(R/k,\tau)}$ obviously maps the homology class of $\Sigma r_0^{(i)} \otimes r_1^{(i)} \otimes \cdots \otimes r_n^{(i)}$ to $\Sigma r_0^{(i)} dr_1^{(i)} \cdot \ldots \cdot dr_n^{(i)}$. Therefore it holds:

$$
\begin{aligned}
&\overline{\delta}_{(R/k,\tau)} \circ \theta_{(R/k,\tau)}^{\cdot} (r_0 dr_1 \cdot \ldots \cdot dr_n) \\
&= \overline{\delta}_{(R/k,\tau)} \left(\text{homology class of} \sum_{\sigma \in S_n} \text{sign} (\sigma) r_0 \otimes r_{\sigma(1)} \otimes \cdots \otimes r_{\sigma(n)} \right) \\
&= \sum_{\sigma \in S_n} \text{sign} (\sigma) r_0 dr_{\sigma(1)} \cdot \ldots \cdot dr_{\sigma(n)} \\
&= n! r_0 dr_1 \cdot \ldots \cdot dr_n,
\end{aligned}
$$

implying i) in this case.

In the general case there exists a commutative diagram

$$
\begin{array}{ccc}
H_{\cdot}\left(\hat{R}/k\right) & \xrightarrow{\ \overline{\delta}_{\hat{R}/k}\ } & \Omega_{\hat{R}/k}^{\cdot} \\
\downarrow{\varphi} & & \downarrow{\psi} \\
H_{\cdot}\left(\hat{R}/k, \tau\right) & \xrightarrow{\ \overline{\delta}_{(R/k,\tau)}\ } & \Omega_{(R/k,\tau)}^{\cdot}
\end{array}
$$

with φ and ψ being the canonical maps. From this and from the above we get for $\omega \in \Omega^n_{\hat{R}/k}$:

$$n! \psi(\omega) = \overline{\delta}_{(R/k,\tau)} \circ \theta_{(R/k,\tau)}(\psi(\omega)).$$

Being a homomorphism of differential algebras of $\hat{R}/k$, ψ is surjective, and the claim follows.

ii) If $\delta'_U : \gamma.(R/U/k) \to \gamma.(R/U/k)$ are the morphisms defining $\Delta_{R/k}$, as constructed in (1.11), then it holds:

$$f^{n+1}_U \circ \delta^n_U = (n+1) d_{R/U/k} \circ f^n_U \quad (n \in \mathbf{N}, U \in \mathfrak{A}).$$

Passing to the projective limit and to homology yields ii).

iii) We only need to show iii) for elements of the form $\omega = dr \left(r \in \hat{R} \right)$. Denoting by $\bar{r}$ the residue class of r in $R/U(U \in \mathfrak{A})$ and by $\mu_{1 \otimes \bar{r}} : \gamma.(R/U/k) \to \gamma.(R/U/k)$ the morphism "multiplication by $1 \otimes \bar{r}$", it suffices to show:

$$f^{n+1}_U \circ \mu_{1 \otimes \bar{r}} | \gamma_n(R/U/k) = (n+1) d\bar{r} \cdot f^n_U \text{ for all } n \in \mathbf{N}, U \in \mathfrak{A}.$$

This can be shown by an easy calculation.

2.18. COROLLARY. $\theta^1_{(R/k,\tau)}$ always is injective, having $\overline{\delta}^1_{(R/k,\tau)}$ as a left inverse.

If τ is the discrete topology on R, or if $(R,\tau)/(k,\tau_0)$ satisfies the conditions of (2.9), then $\theta^1_{(R/k,\tau)}$ is bijective with inverse $\overline{\delta}^1_{(R/k,\tau)}$.

Proof: If τ is the discrete topology on R, then it also can be shown easily that $\theta^1_{(R/k,\tau)} \circ \overline{\delta}^1_{(R/k,\tau)} = id_{H_1(R/k,\tau)}$.

If the conditions of (2.9) are satisfied then there exists an $\hat{R}$–isomorphism

$$\Omega^1_{(R/k,\tau)} \xrightarrow{\sim} H_1(R/k,\tau).$$

Since $\overline{\delta}^1_{(R/k,\tau)}$ is surjective, and since $\Omega^1_{(R/k,\tau)}$ is a noetherian $\hat{R}$-module, $\overline{\delta}^1_{(R/k,\tau)}$ is bijective with inverse $\theta^1_{(R/k,\tau)}$.

2.19. COROLLARY. *If in addition* $\mathbb{Q} \subseteq R$, *then there exists a functorial homomorphism of graded* $\hat{R}$-*modules*

$$\delta'_{(R/k,\tau)} : H.(R/k,\tau) \to \Omega'_{(R/k,\tau)}$$

satisfying:

i) $\delta'_{(R/k,\tau)} \circ \theta'_{(R/k,\tau)} = id_{\Omega'_{(R/k,\tau)}}.$
In particular $\theta'_{(R/k,\tau)}$ *is injective*

ii) $d_{R/k} \circ \delta'_{(R/k,\tau)} = \delta'_{(R/k,\tau)} \circ \Delta_{R/k}$

iii) $\delta'_{(R/k,\tau)} \left(\theta'_{(R/k,\tau)}(\omega) \cdot x \right) = \omega \cdot \delta'_{(R/k,\tau)}(x)$ *for* $\omega \in \Omega'_{(R/k,\tau)}, x \in H.(R/k,\tau)$, *i.e.* $\delta'_{(R/k,\tau)}$ *is* $\Omega'_{(R/k,\tau)}$-*linear, if we consider* $H.(R/k,\tau)$ *as a left* $\Omega'_{(R/k,\tau)}$-*module via* $\theta'_{(R/k,\tau)}$.

Proof: $\delta'_{(R/k,\tau)}$, defined by $\delta^n_{(R/k,\tau)} := \frac{1}{n!} \bar{\delta}^n_{(R/k,\tau)}$ is $\hat{R}$-linear and has the desired properties by (2.17).

Let $\Omega' = \Lambda_R (\Omega^1)$ be an arbitrary exterior differential algebra of R/k such that Ω^1 is a finite free R-module. In this case Hochschild homology and Ω' are also closely related ([**L₁**], (4.6.4)):

For an R-bimodule M let

$$H^n(R/k,M) := H_n \left(Hom_{R^e}(\beta.(R/k), M) \right)$$

be the nth Hochschild cohomology module of M ([**ML**], X.3). It is well known ([**ML**], X.(3.4)) that

$$H^1(R/k,M) = \text{Der}_k(R,M)/\{\text{inner derivations from } R \text{ to } M\}$$

In particular if M is an R-module, considered as an R-bimodule in the canonical way, this becomes

$$H^1(R/k,M) = \text{Der}_k(R,M).$$

Let $\omega_1, \ldots, \omega_r$ be a basis of Ω^1 and let $\omega_1^*, \ldots, \omega_r^*$ be the corresponding dual basis of $\text{Hom}_R(\Omega^1, R)$. If $d : R \to \Omega^1$ is the k-derivation induced by the differentiation of Ω', the maps $d_i := \omega_i^* \circ d : R \to R$ are k-derivations. They induce elements $[d_i] \in H^1(R/k,R)(i = 1, \ldots, r)$. Since $H'(R/k,R) := \bigoplus_{n \in \mathbb{N}} H^n(R/k,R)$ has a canonical structure of graded associative R-algebra ([**L₁**], (1.8)), there exist elements

$$\delta^{(n)} := \sum_{i_1 < i_2 < \cdots < i_n} [d_{i_1}] \cdot [d_{i_2}] \cdot \ldots \cdot [d_{i_n}] \otimes \omega_{i_1} \cdot \omega_{i_2} \cdot \ldots \cdot \omega_{i_n} \in H^n(R/k,R) \otimes \Omega^n = H^n(R/k,\Omega^n).$$

2.20. Remark: $2\delta^{(n)}$ is independent of the choice of a basis of Ω^1.

The proof of this remark is based on the fact that for $d_1, d_2 \in H^1(R/k, R)$, $d_1 \cdot d_2 + d_2 \cdot d_1 = 0$ in the ring $H^{\cdot}(R/k, R)$.

By [L_1], (1.1) there exists a canonical pairing

$$H^n(R/k, \Omega^n) \otimes_R H_n(R/k) \to \Omega^n \qquad (n \in \mathbf{N})$$

Therefore the $\delta^{(n)}$ induce morphisms

$$\delta^n_{R/k} : H_n(R/k) \to \Omega^n \qquad (n \in \mathbf{N})$$

These maps can be described as follows:

If $\sum_j r_0^{(j)} \otimes r_1^{(j)} \otimes \cdots \otimes r_n^{(j)} \in \gamma_n(R/k)$ is an n–cycle representing an element $x \in H_n(R/k)$, then

$$\delta^n_{R/k}(x) = \sum_j \sum_{i_1 < i_2 < \cdots < i_n} r_0^{(j)} d_{i_1}\left(r_1^{(j)}\right) \cdot d_{i_2}\left(r_2^{(j)}\right) \cdot \ldots \cdot d_{i_n}\left(r_n^{(j)}\right) \cdot \omega_{i_1} \cdot \omega_{i_2} \cdot \ldots \cdot \omega_{i_n}.$$

2.21. PROPOSITION.

i) For every $n \in \mathbf{N}$ there exist commutative diagrams

$$\Omega^n_{R/k} \xrightarrow{\theta^n_{R/k}} H_n(R/k) \text{ and } H_n(R/k) \xrightarrow{\delta^n_{R/k}} \Omega^n_{R/k}$$

with φ being the canonical map. In particular $\delta^n_{R/k}$ coincides with the map $\delta^n_{R/k}$ defined in (2.19), if $\mathbf{Q} \subseteq R$ and $\Omega^{\cdot} = \Omega^{\cdot}_{R/k}$.

ii) If 2 is not a zero–divisor on R, then $\delta^n_{R/k}$ is independent of the choice of a basis of Ω^1.

iii) $\delta_{R/k}\left(\theta^{\cdot}_{R/k}(\omega) \cdot x\right) = \varphi(\omega) \cdot \delta_{R/k}(x)$ for $x \in H_{\cdot}(R/k)$ and $\omega \in \Omega^{\cdot}_{R/k}$.

Proof:

i) For $r_0 dr_1 \cdot \ldots \cdot dr_n \in \Omega^n_{R/k}$ it holds:

$$\delta^n_{R/k}\left(\theta^n_{R/k}(r_0 dr_1 \cdot \ldots \cdot dr_n)\right)$$

$$= \delta^n_{R/k}\left(\text{homology class of } \sum_{\sigma \in S_n} \text{sign}(\sigma) r_0 \otimes r_{\sigma(1)} \otimes \cdots \otimes r_{\sigma(n)}\right)$$

$$= \sum_{i_1 < \ldots, < i_n} \sum_{\sigma \in S_n} \text{sign}(\sigma) r_0 d_{i_1}\left(r_{\sigma(1)}\right) \cdot \ldots \cdot d_{i_n}\left(r_{\sigma(n)}\right) \cdot \omega_{i_1} \cdot \ldots \cdot \omega_{i_n}$$

$$= \varphi(r_0 dr_1 \cdot \ldots \cdot dr_n).$$

If $[d] \in H^1\left(R/k, \Omega^1\right)$ is the element induced by $d : R \to \Omega^1$, then it holds: $n! \delta^{(n)} = [d]^n$ in $H^n\left(R/k, \Omega^n\right)$. For the map $H_n(R/k) \to \Omega^n$ induced by $[d]^n$ the commutativity of the second diagram can be verified easily.

ii) is an immediate consequence of (2.20).

iii) has to be shown only for elements of the form $\omega = dr (r \in R)$. This can be done by an easy calculation.

I do not know if $\delta^n_{R/k}$ is in general independent of the choice of a basis of Ω^1, even if 2 is a zero–divisor on R. It can be shown in the following case:

2.22. PROPOSITION. *If R is noetherian and reduced, then*

$$\delta_{R/\mathbf{Z}} : H_{\cdot}(R/\mathbf{Z}) \to \Omega^{\cdot}$$

is independent of the choice of a basis of Ω^1.

Proof: Let $\mathfrak{A}$ be a basis of Ω^1, and denote by $\delta^{\mathfrak{A}}_{R/\mathbf{Z}}$ the morphism constructed using $\mathfrak{A}$. Let $K := Q(R)$ be the full ring of quotients of R. $\mathfrak{A}$ defines in a canonical way a basis of Ω^1_K which again will be denoted by $\mathfrak{A}$. The explicit description of $\delta_{R/\mathbf{Z}}$ shows that the diagram

$$
\begin{array}{ccc}
H_{\cdot}(R/\mathbf{Z}) & \xrightarrow{\;\delta^{\mathfrak{A}}_{R/\mathbf{Z}}\;} & \Omega^{\cdot} \\
\downarrow & & \downarrow \\
H_{\cdot}(K/\mathbf{Z}) & \xrightarrow{\;\delta^{\mathfrak{A}}_{K/\mathbf{Z}}\;} & \Omega^{\cdot}_K
\end{array}
$$

commutes. Since $\Omega^{\cdot} \to \Omega^{\cdot}_K$ is injective, it suffices to show that $\delta^{\mathfrak{A}}_{K/\mathbf{Z}}$ is independent of the choice of a basis. K is a finite direct product of fields, and by (2.21)i) the following diagram commutes:

$$
\begin{array}{ccc}
\Omega^{\cdot}_{K/\mathbf{Z}} & \xrightarrow{\;\theta_{K/\mathbf{Z}}\;} & H_{\cdot}(K/\mathbf{Z}) \\
 & \searrow \qquad \swarrow \delta^{\mathfrak{A}}_{K/\mathbf{Z}} & \\
 & \Omega^{\cdot}_K &
\end{array}
$$

Therefore the claim follows from:

2.23. LEMMA. *If R is a finite direct product of fields, then*

$$\theta_{R/\mathbf{Z}} : \Omega^{\cdot}_{R/\mathbf{Z}} \to H_{\cdot}(R/\mathbf{Z})$$

is an isomorphism of DG–algebras.

Proof: By (1.18), [**KD**], (4.7) and the functoriality of $\theta^{.}$ we may assume: R is a field. If k is the prime field of R, then the canonical maps $\Omega^{.}_{R/\mathbb{Z}} \to \Omega^{.}_{R/k}$ and $H.(R/\mathbb{Z}) \to H.(R/k)$ are isomorphisms. Since R/k is separable, $\theta^{.}_{R/k}$ is an isomorphism by (2.13), and the claim follows.

§3. Traces in Hochschild homology.

The canonical trace ([**KD**], appendix F)

$$\sigma_{S/R} : S \to R$$

of a finite projective algebra S can be generalized to Hochschild homology ([**L$_1$**], (4.5)). In this section traces will be constructed also for the Hochschild homology of topological algebras. The method we will use is more concrete than the approach in [**L$_1$**]. For discretely topologized algebras it was suggested to us by J. Lipman (private communication).

Let X be the class of all triples $((S,\tau'),(R,\tau),k)$ having the following properties:

i) k is a commutative ring and $(R,\tau)/k$ is a topological algebra. (We think of k as being equipped with the discrete topology.)

ii) S/R is an algebra, and the topology τ' on S is the linear topology on S induced by τ, i.e.: If $\mathfrak{U}$ is a fundamental set of neighborhoods of $0 \in R$, consisting of ideals on R, then $\mathfrak{U}' := \{US : U \in \mathfrak{U}\}$ is a fundamental set of neighborhoods of $0 \in S$, consisting of ideals of S, for the topology τ'.

iii) The completion $\left(\hat{S},\widehat{\tau'}\right)$ of (S,τ) is a finite and free module over the completion $\left(\hat{R},\hat{\tau}\right)$ of (R,τ).

iv) The topology $\widehat{\tau'}$ on $\hat{S}$ is the linear topology on $\hat{S}$ induced by $\hat{\tau}$ (as defined in ii)).

3.1. Remark: Condition ii) implies iv) if S/R is finite ([**B$_1$**], III. §12.2, Prop. 16), or if τ is the $\mathfrak{J}$–adic topology on R for some finitely generated ideal $\mathfrak{J} \subset R$ ([**N**], thm. 17.4). If τ is the discrete topology on R, then ii)–iv) hold if and only if S/R is a finite free algebra.

Denote by $M_r(R)$ the ring of $r \times r$–matrices with coefficients in R. There exists a canonical morphism

$$Sp : T_k^{n+1}\left(M_r(R)\right) \to T_k^{n+1}(R) = \gamma_n(R/k)$$

(notations as in §1), defined by

$$Sp\left(A^0 \otimes A^1 \otimes \cdots \otimes A^n\right) = \sum_{i_0,i_1,\ldots,i_n=1} A^0_{i_0,i_1} \otimes A^1_{i_1,i_2} \otimes \cdots \otimes A^n_{i_n,i_0}$$

for $r \times r$–matrices $A^\rho = \left(A^\rho_{i,j}\right)_{i,j=1,\ldots,r}, \rho = 0,\ldots,n.$

Let S/R be a finite free algebra with basis $\mathfrak{A}$ and with $\operatorname{rank}_R(S) = r$. By $\rho_{\mathfrak{A}}$ we denote the regular representation of S/R with respect to the basis $\mathfrak{A}$:

$$\rho_{\mathfrak{A}} : S \to M_r(R), s \mapsto Ma^{\mathfrak{A}}(\mu_s),$$

where $\mu_s \in \operatorname{Hom}_R(S, S)$ is the homomorphism "multiplication by s", and where $Ma^{\mathfrak{A}}(\mu_s)$ is its matrix with respect to $\mathfrak{A}$.

Now let $((S, \tau'), (R, \tau), k)$ be in X, let $\mathfrak{U}$ be a fundamental set of neighborhoods of $0 \in R$, consisting of ideals of R, and let $\overline{\mathfrak{U}}$ be the corresponding fundamental set of neighborhoods of $0 \in \hat{R}$. We may assume: $\hat{R} \neq (0)$.

If $\mathfrak{A}$ is an $\hat{R}$–basis of $\hat{S}$, then $\mathfrak{A}$ induces an $\hat{R}/\overline{U} = R/U$–basis of $\hat{S}/\overline{U}\hat{S} = S/US$ for every $U \in \mathfrak{U}$. The trace homomorphism

$$Sp^{\mathfrak{A}} : \gamma. \left(\hat{S}/k \right) \to \gamma. \left(\hat{R}/k \right),$$

defined by

$$Sp^{\mathfrak{A}}(s_0 \otimes s_1 \otimes \cdots \otimes s_n) := Sp(\rho_{\mathfrak{A}}(s_0) \otimes \rho_{\mathfrak{A}}(s_1) \otimes \cdots \otimes \rho_{\mathfrak{A}}(s_n)) \left(s_0, s_1, \ldots, s_n \in \hat{S}, n \in \mathbb{N} \right)$$

is a morphism of complexes and induces a projective system of complex homomorphisms

$$Sp^{\mathfrak{A}v} : \gamma.(S/US/k) \to \gamma.(R/U/k).$$

Using condition ii) we get a well defined homomorphism of complexes

$$\widetilde{Sp}^{\mathfrak{A}} : \gamma.(S/k, \tau') \to \gamma.(R/k, \tau)$$

of the topological Hochschild complexes, making the following diagram commutative:

$$
\begin{array}{ccc}
\gamma. \left(\hat{S}/k \right) & \xrightarrow{\ Sp^{\mathfrak{A}}\ } & \gamma. \left(\hat{R}/k \right) \\
\Big\downarrow {\scriptstyle \text{can.}} & & \Big\downarrow {\scriptstyle \text{can.}} \\
\gamma.(S/k, \tau') & \xrightarrow{\ \widetilde{Sp}^{\mathfrak{A}}\ } & \gamma.(R/k, \tau),
\end{array}
$$

and inducing a well defined map

$$tr^{\mathfrak{A}}_{S/R} : H.(S/k, \tau') \to H.(R/k, \tau)$$

of topological Hochschild homology groups.

3.2. LEMMA. $tr^{\mathfrak{A}}_{S/R}$ is independent of the choice of a basis $\mathfrak{A}$ of $\hat{S}/\hat{R}$.

Proof: Let $\mathfrak{A}$ and $\mathfrak{B}$ be $\hat{R}$–basis of $\hat{S}$ and let $\rho_{\mathfrak{A}} : \hat{S} \to M_r\left(\hat{R}\right)$ and $\rho_{\mathfrak{B}} : \hat{S} \to M_r\left(\hat{R}\right)$ be the corresponding regular representations. Then there exists an invertible matrix $A \in M_r(R)$ such that $\rho_{\mathfrak{A}}(s) = A \cdot \rho_{\mathfrak{B}}(s) \cdot A^{-1}$ for all $s \in \hat{S}$.

The k–linear maps

$$\alpha_n : \gamma_n\left(\hat{S}/k\right) \to \gamma_{n+1}\left(\hat{R}/k\right),$$

defined by

$$\alpha_n\left(s_0 \otimes s_1 \otimes \cdots \otimes s_n\right)$$
$$= Sp\left(A \otimes \rho_{\mathfrak{B}}\left(s_0\right)A^{-1} \otimes A\rho_{\mathfrak{B}}\left(s_1\right)A^{-1} \otimes \cdots \otimes A\rho_{\mathfrak{B}}\left(s_n\right)A^{-1}\right)$$
$$+ Sp\left(\sum_{i=0}^{n}(-1)^{in}A \otimes \rho_{\mathfrak{B}}\left(s_i\right) \otimes \rho_{\mathfrak{B}}\left(s_{i+1}\right) \otimes \cdots \otimes \rho_{\mathfrak{B}}\left(s_n\right)\right.$$
$$\left.\otimes \rho_{\mathfrak{B}}\left(s_0\right)A^{-1} \otimes A\rho_{\mathfrak{B}}\left(s_1\right)A^{-1} \otimes \cdots \otimes A\rho_{\mathfrak{B}}\left(s_{i-1}\right)A^{-1}\right)$$

induce a projective system of morphisms

$$\alpha_n^U : \gamma_n(S/US/k) \to \gamma_{n+1}(R/U/k)$$

satisfying

$$\alpha_{n-1}^U \circ d_n + d_{n+1} \circ \alpha_n^U = Sp_n^{\mathfrak{A}_U} - Sp_n^{\mathfrak{B}_U}$$

$$(n \in \mathbb{N}, U \subseteq R \text{ an open ideal}).$$

Passing to the projective limit shows that $\widetilde{Sp}^{\mathfrak{A}}$ and $\widetilde{Sp}^{\mathfrak{B}}$ are homotopic, implying

$$tr^{\mathfrak{A}}_{S/R} = tr^{\mathfrak{B}}_{S/R}.$$

In the following we will write $tr^k_{(S/R,\tau)}$ instead of $tr^{\mathfrak{A}}_{S/R}$, or simply $tr^k_{S/R}$ if τ is the discrete topology. If no confusion is likely we also will omit the superscript k.

3.3. Remark: There exists a commutative diagram

$$
\begin{array}{ccccc}
H.\left(\hat{S}/k\right) & \longrightarrow & H.\left(\hat{S}/k, \widehat{\tau'}\right) & \xleftarrow{\ \sim\ } & H.(S/k,\tau) \\
\downarrow{\scriptstyle tr_{S/R}} & & \downarrow{\scriptstyle tr_{(S/R,+)}} & & \downarrow{\scriptstyle tr_{(S/R,\tau)}} \\
H.\left(\hat{R}/k\right) & \longrightarrow & H.\left(\hat{R}/k, \hat{\tau}\right) & \xleftarrow{\ \sim\ } & H.(R/k,\tau),
\end{array}
$$

where unlabelled arrows represent natural maps.

Next we will prove the theorem mentioned in the introduction about the traces $tr_{(S/R,\tau)}$.

3.4. THEOREM. *The system* $\left\{ tr^k_{(S/R,\tau)} : ((S,\tau'),(R,\tau),k) \text{ in } X \right\}$ *satisfies the following trace axioms:*

$$(T1) \qquad (H.(R/k,\tau)\text{–linearity})$$

If we consider $H.(S/k,\tau')$ *as a left* $H.(R/k,\tau)$*–module via the canonical map*

$$H.(R/k,\tau) \to H.(S/k,\tau'),$$

then $tr_{(S/R,\tau)}$ *is* $H.(R/k,\tau)$*–linear and homogeneous of degree 0.*

$$(T2) \qquad (\text{Relation to the canonical trace})$$

The restriction $tr_{(S/R,\tau)}\big| H_0(S/k,\tau') : \hat{S} \to \hat{R}$ *to the elements of degree 0 is the canonical trace* $\sigma_{\hat{S}/\hat{R}}$.

$$(T3) \qquad (\text{Base change})$$

Let $(R,\tau)/k \to (R',\tau_1)/k'$ *be a morphism of topological algebras, let* $S' := S \otimes_R R'$ *and let* τ_1' *be the linear topology on* S' *induced by* τ_1. *If* $((S',\tau_1'),(R',\tau_1),k')$ *is in* X, *then the following diagram commutes:*

$$
\begin{array}{ccc}
H.(S/k,\tau') & \longrightarrow & H.(S'/k',\tau_1') \\
\downarrow {\scriptstyle tr_{(S/R,\tau)}} & {\scriptstyle tr_{(S'/R',\tau_1)}} \downarrow & \\
H.(R/k,\tau) & \longrightarrow & H.(R'/k',\tau_1).
\end{array}
$$

$$(T4) \qquad (\text{direct products})$$

Let $S = S_1 \times \cdots \times S_h$ *be a direct product of* R*–algebras, and suppose that also* $((S_1,\tau_1'),(R,\tau),k),\ldots,((S_h,\tau_h'),(R,\tau),k)$ *are in* X *(with* τ_i' *being the linear topology on* S_i *induced by* τ*). Then the following diagram commutes:*

$$
\begin{array}{ccc}
H.(S/k,\tau') & \longrightarrow & H.(S_1/k,\tau_1') \times \cdots \times H.(S_h/k,\tau_h') \\
 & \searrow {\scriptstyle tr_{(S/R,\tau)}} \qquad \sum tr_{(S_i/R,\tau)} \swarrow & \\
 & H.(R/k,\tau). &
\end{array}
$$

$$(T5) \qquad (\text{Transitivity})$$

If $((S,\tau'),(R,\tau),k)$ *and* $((T,\tau''),(S,\tau'),k)$ *are in* X, *then also* $((T,\tau''),(R,\tau),k)$ *is in* X, *and it holds*

$$tr_{(S/R,\tau)} \circ tr_{(T/S,\tau')} = tr_{(T/R,\tau)}.$$

(T6)   (Compatibility with differentiation)

Denote by d_R resp. d_S the differentiation of $H.(R/k, \tau)$ resp. $H.(S/k, \tau')$ as defined in (1.11). Then it holds:

$$d_R \circ tr_{(S/R,\tau)} = tr_{(S/R,\tau)} \circ d_S.$$

(T7)   (Logarithmic derivative)

Let $n_{\hat{S}/\hat{R}} : \hat{S} \to \hat{R}$ be the canonical norm of $\hat{S}/\hat{R}$ ([KD], app. F). If $a \in \hat{S}$ is a unit, then it holds:

$$tr_{(S/R,\tau)} \left(\frac{d_S(a)}{a} \right) = \frac{d_R \left(n_{\hat{S}/\hat{R}}(a) \right)}{n_{\hat{S}/\hat{R}}(a)}.$$

(T8)   (local algebras)

Let $\hat{S}/\hat{R}$ be a local algebra, let $m_{\hat{S}}$ resp. $m_{\hat{R}}$ be the maximal ideal of $\hat{S}$ resp. $\hat{R}$ and let $L := \hat{S}/m_{\hat{S}}, K := \hat{R}/m_{\hat{R}}$. Let $n := \mathrm{rank}_{\hat{R}} \left(\hat{S} \right) \cdot \dim_K(L)^{-1}$. Then the following diagram commutes:

$$
\begin{array}{ccc}
H.(S/k, \tau') & \longrightarrow & H.(L/k) \\
\downarrow{\scriptstyle tr_{(S/R,\tau)}} & {\scriptstyle n \cdot tr_{L/K}} \downarrow & \\
H.(R/k, \tau) & \longrightarrow & H.(K/k).
\end{array}
$$

The horizontal maps are induced by the canonical epimorphisms which are continuous (if we exclude the case $\hat{R} = (0)$)..

Proof: The axioms (T1)–(T8) will be shown in the discrete case first. Then the general case will be reduced to this case.

a) Let τ be the discrete topology on R.

(T1): Recall that $\gamma.(R/k)$ and $\gamma.(S/k)$ carry canonical algebra structures ((1.3)i). If we consider $\gamma.(S/k)$ as a left $\gamma.(R/k)$–module via the canonical map $\gamma.(R/k) \to \gamma.(S/k)$, and if $\mathfrak{A}$ is an R–basis of S, then

$$Sp^{\mathfrak{A}} : \gamma.(S/k) \to \gamma.(R/k)$$

is $\gamma.(R/k)$–linear and homogeneous of degree 0. From this (T1) follows.

(T2) is evident.

(T3): If $\mathfrak{A}$ is an R–basis of S, then $\mathfrak{A}' := \{1 \otimes a : a \in \mathfrak{A}\}$ is an R'–basis of S'. A trivial calculation shows that

$$
\begin{array}{ccc}
\gamma.(S/k) & \xrightarrow{\text{can.}} & \gamma.(S'/k') \\
\downarrow {\scriptstyle Sp^{\mathfrak{A}}} & & \downarrow {\scriptstyle Sp^{\mathfrak{A}'}} \\
\gamma.(R/k) & \xrightarrow{\text{can.}} & \gamma.(R'/k')
\end{array}
$$

commutes, proving (T3).

(T4): If $\mathfrak{A}_i = \left\{s_1^i, \ldots, s_{r_i}^i\right\}$ are R–bases of $S_i (i = 1, \ldots, h)$, then $\mathfrak{A} = \left\{s_1^1, \ldots, s_{r_1}^1, s_1^2, \ldots, s_{r_h}^h\right\}$ is an R–basis of S, and the following diagram commutes:

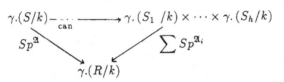

This implies (T4).

(T5): If $\mathfrak{A} = \{y_1, \ldots, y_n\}$ is an R–basis of S, and if $\mathfrak{A}_2 = \{z_1, \ldots, z_m\}$ is an S–basis of T, then $\mathfrak{A} = \{y_1 z_1, y_2 z_1, \ldots, y_n z_m\}$ is an R–basis of T. Let $\rho_{\mathfrak{A}_1} : S \to M_n(R), \rho_{\mathfrak{A}_2} : T \to M_m(S)$ and $\rho_{\mathfrak{A}} : T \to M_{nm}(R)$ be the corresponding regular representations. For $t \in T$ let $\rho_{\mathfrak{A}_2}(t) = (s_{i,j})$. Then $\rho_{\mathfrak{A}}(t) = (B_{i,j})$ with $B_{i,j} = \rho_{\mathfrak{A}_1}(s_{i,j})$. From this $Sp^{\mathfrak{A}} = Sp^{\mathfrak{A}_1} \circ Sp^{\mathfrak{A}_2}$ follows easily, hence

$$
tr_{T/R} = tr_{S/R} \circ tr_{T/S}.
$$

(T6): Let $\delta^R : \gamma.(R/k) \to \gamma.(R/k)$ resp. $\delta^S : \gamma.(S/k) \to \gamma.(S/k)$ be the maps inducing d_R resp. d_S (see (1.3)ii)). If $\mathfrak{A}$ is a basis of S, then an easy calculation shows

$$
Sp^{\mathfrak{A}} \circ \delta^S = \delta^R \circ Sp^{\mathfrak{A}},
$$

implying (T6).

(T7): Let $\mathfrak{A}$ be an R–basis of S and let $B := \rho_{\mathfrak{A}}(a)$. Furthermore let $A = (A_{i,\ell})$ be the adjoint of B. Then $B^{-1} = \det(B)^{-1} A = n_{S/R}(a)^{-1} A$, and we get for the differentiation d_R of $H.(R/k)$:

$$
d_R(\det(B)) = \sum_{\ell, i=1}^{r} A_{\ell, i} d_R(B_{i,\ell}) \qquad (\textbf{[KD]}, (1.9)\text{h}).
$$

On the other hand it holds:

$$
Sp^{\mathfrak{A}} \left(a^{-1} \cdot 1 \otimes a\right) = Sp \left(B^{-1} \cdot 1 \otimes B\right)
$$

$$
= n_{S/R}(a)^{-1} \sum_{i, \ell=1}^{r} A_{\ell, i} \cdot \delta^R(B_{i, \ell})
$$

(with δ^R as in the proof of (T6)). From this the claim follows.

(T8): By (T3) it suffices to show the claim for R/m_R resp. $S/m_R S$ instead of R resp. S. So we may assume: $R = K$.

i) Let $B^0, B^1, \ldots, B^m \in M_r(R), \nu \in \mathbb{N}$ such that $\nu | r$, let $n := \frac{r}{\nu}$ and suppose that

$$
B^i = \begin{pmatrix} A^i & & & & \\ & A^i & & * & \\ & & \ddots & & \\ & 0 & & & \\ & & & & A^i \end{pmatrix} \qquad \left(A^i \in M_\nu(R) \right).
$$

Then it obviously holds:

$$
Sp\left(B^0 \otimes B^1 \otimes \cdots \otimes B^m \right) = n \cdot Sp\left(A^0 \otimes A^1 \otimes \cdots \otimes A^m \right)
$$

ii) Let $x_1, \ldots, x_\nu$ be elements of S whose residue classes $\bar{x}_1, \ldots, \bar{x}_\nu$ in L form an R-basis of L, and let $\rho : L \to M_\nu(R)$ be the corresponding regular representation of L.

Since $\dim(S) = 0$ there exists an $\alpha \in \mathbb{N}$ such that $m_S^\alpha = 0$. For $i \in \{1, \ldots, \alpha - 1\}$ let $\left\{ y_1^{(i)}, \ldots, y_{\nu_i}^{(i)} \right\}$, be a system of representatives of an L-basis of m_S^i/m_S^{i+1} in m_S^i. Then

$$
\mathfrak{A} := \{x_1, \ldots, x_\nu\} \cup \underbrace{\bigcup_{i=1}^{\alpha-1} \left\{ x_j y_\ell^{(i)} : j \in \{1, \ldots, \nu\}, \ell \in \{1, \ldots, \nu_i\} \right\}}_{=:\{x_{\nu+1}, \ldots, x_{n\nu}\}}
$$

is an R-basis of S.

If $\rho_{\mathfrak{A}} : S \to M_{\nu n}(R)$ is the corresponding regular representation of S/R, then we get:

$$
\rho_{\mathfrak{A}}(x_i) = \begin{pmatrix} \rho(\bar{x}_i) & & & & \\ & \rho(\bar{x}_i) & & * & \\ & & \ddots & & \\ & 0 & & & \\ & & & & \rho(\bar{x}_i) \end{pmatrix} \qquad (i = 1, \ldots, \nu)
$$

(here $^-$ denotes the residue class of an element in L), and

$$
\rho_{\mathfrak{A}}(x_i) = \begin{pmatrix} 0 & & & & \\ & 0 & & * & \\ & & \ddots & & \\ & 0 & & & \\ & & & & 0 \end{pmatrix} \qquad (i = \nu + 1, \ldots, \nu n).
$$

Since it suffices to show the claim on the level of the defining complexes for elements of the form $r_0 x_{i_0} \otimes r_1 x_{i_1} \otimes \cdots \otimes r_n x_{i_n}$ $(i_0, \ldots, i_n \in \{1, \ldots, n\nu\})$ (T8) follows from the above and i).

b) General case $(WLOG : \hat{R} \neq (0))$:

(T1)–(T6) follow easily from the corresponding proofs in a), since these properties already hold for the systems of complex morphisms which define the trace (suitable bases chosen).

As for (T3) we also have to show: $\hat{S}' = \widehat{R'} \otimes_{\hat{R}} \hat{S}$. It obviously suffices to show:

If $\mathfrak{A}$ is an $\hat{R}$–basis of $\hat{S}$, then the image $\mathfrak{A}'$ of $\mathfrak{A}$ by $\hat{S} \xrightarrow{\text{can}} \widehat{R'} \otimes_{\hat{R}} \hat{S} \xrightarrow{\text{can}} \hat{S}'$ is an $\widehat{R'}$–basis of $\hat{S}'$.

Let $r := \operatorname{rank}_{\hat{R}}\left(\hat{S}\right)$. For every open ideal $U \subseteq R$ the R/U–basis $\mathfrak{A}^U$ of S/US induces in a canonical way an R'/UR'–basis $\mathfrak{A}'^U$ of $S'/US' = S/US \otimes_{R/U} R'/UR'$. By assumption there exists for every open ideal $V \subseteq R'$ an open ideal $U \subseteq R$ such that $UR' \subseteq V$. Therefore $\mathfrak{A}'^U$ induces an $R'/V = \widehat{R'}/\overline{V}$–basis $\mathfrak{A}'^V$ of $\hat{S}'/\overline{V}\hat{S}' = S'/VS' = S'/US' \otimes_{R'/UR'} R'/V$. Furthermore $\mathfrak{A}'^V$ is the image of $\mathfrak{A}'$ by $\hat{S}' \to \hat{S}'/\overline{V}\hat{S}'$. Therefore $\mathfrak{A}'$ induces a projective system of isomorphisms

$$(R'/V)^r \to S'/VS'$$

of R'/V–modules $(V \subseteq R'$ an open ideal), and consequently $\mathfrak{A}'$ is an $\hat{R}'$–basis of $\hat{S}'$.

As for (T4) it remains to remark that $\hat{S} = \hat{S}_1 \times \cdots \times \hat{S}_h$ since the functor "completion with respect to the topology τ" commutes with finite direct sums and products.

(T7) follows from a) by the commutativity of the diagram

$$
\begin{CD}
H.\left(\hat{S}/k\right) @>\psi>> H.(S/k, \tau') \\
@V{trs/R}VV @VV{tr_{(S/R, \tau)}}V \\
H.\left(\hat{R}/k\right) @>\varphi>> H.(R/k, \tau)
\end{CD}
$$

and by the fact that $\frac{ds(a)}{a} \in im(\psi)$, since ψ and φ are morphisms of DG–algebras.

(T8): By (T3) we may replace R by K. Since the topology induced by τ on K is the discrete one, (T8) follows from a). ∎

3.5. Remark: Let R/k be flat and let S/R be finite and projective. Furthermore let $\mathcal{F}$ be the family of all $f \in R$ such that S_f is a free R_f–module. Then we have $\operatorname{Spec}(S) =$

$\underset{f\in\mathcal{F}}{\cup}D(f)$. Since also S/k is flat there exist canonical isomorphisms $H.(S_f/k) \cong H.(S/k)_f$ and $H.(R_f/k) \cong H.(R/k)_f$ for every $f \in \mathcal{F}$ ((1.15)).

Therefore the trace morphisms define maps

$$tr_{S_f/R_f} : H.(S/k)_f \to H.(R/k)_f.$$

It follows immediately from (T3) and (1.15) that $\left(tr_{S_f/R_f}\right)_g = tr_{S_{fg}/R_{fg}}$ for $f,g \in \mathcal{F}$, and therefore there exists a unique R–linear map

$$tr_{S/R} : H.(S/k) \to H.(R/k)$$

such that $\left(tr_{S/R}\right)_f = tr_{S_f/R_f}$ for all $f \in \mathcal{F}$.

By construction these maps satisfy the trace axioms with the possible exception of (T5).

3.6. Remark: Let R/k be an arbitrary algebra and let S/R be finite and projective. Then Lipman defines in $[\mathbf{L_1}]$, (4.5) a trace morphism

$$tr_{S/R} : H.(S/k) \to H.(R/k).$$

It is easy to see that this is the trace constructed above if S/R is free, or if R/k is flat. Making the obvious choices it can be seen that the trace of $[\mathbf{L_1}]$, (4.5) is induced by a complex homomorphism $\gamma.(S/k) \to \gamma.(R/k)$ which is continuous, if $(R,\tau)/k$ is a topological algebra, hence induces

$$tr_{(S/R,\tau)} : H.(S/k,\tau') \to H.(R/k,\tau).$$

We will freely use this remark to talk of traces, even if the assumptions of (3.4) are not satisfied.

§4. Traces of Differential Forms.

Under suitable assumptions a theory of traces of differential forms can be derived from the theory of traces in Hochschild homology. This approach was discovered by Lipman ([L$_1$], (4.6)). In particular it provides an intrinsic definition of traces and pretraces as constructed by Angéniol ([A], (7.1.2)), which renders unnecessary all the computations in [A], pp. 109–113.

Let S/R be free. Defining

$$\tau_{S/R}^{\Omega_{R/\mathbf{Z}}} := \bar{\delta}_{R/\mathbf{Z}} \circ tr_{S/R}^{\mathbf{Z}} \circ \theta_{S/\mathbf{Z}}$$

with the maps $\bar{\delta}_{R/\mathbf{Z}}$ and $\theta_{S/\mathbf{Z}}$ of (2.10) and (2.17) we get a well–defined R–linear map

$$\tau_{S/R}^{\Omega_{R/\mathbf{Z}}} : \Omega_{S/\mathbf{Z}} \to \Omega_{R/\mathbf{Z}}.$$

Now let Ω be an arbitrary differential algebra of R. Then $\Omega = \Omega_{R/\mathbf{Z}}/\mathfrak{J}$, where $\mathfrak{J}$ is a homogeneous differentially closed ideal of $\Omega_{R/\mathbf{Z}}$, and consequently $\Omega_S = \Omega_{S/\mathbf{Z}}/\mathfrak{J}\Omega_{S/\mathbf{Z}}$ ([KD], (3.4)). By (2.10), (2.17) and (3.4) (T3) $\tau_{S/R}^{\Omega_{R/\mathbf{Z}}}\left(\mathfrak{J}\Omega_{S/\mathbf{Z}}\right) \subseteq \mathfrak{J}$, and therefore we get a well–defined map

$$\tau_{S/R}^{\Omega} : \Omega_S \to \Omega.$$

Finally let S/R be finite and projective, and let $\mathfrak{F}$ be the family of all $f \in R$ such that S_f/R_f is free. Then there exists a map

$$\tau_{S_f/R_f}^{\Omega_f} : \Omega_{S_f} = (\Omega_S)_f \to \Omega_f = \Omega_{R_f}$$

for every $f \in \mathfrak{F}$. Since the map $tr_{S_f/R_f}^{\mathbf{Z}}$ is independent of the basis chosen for its construction, (3.4) (T3) implies

$$\left(\tau_{S_f/R_f}^{\Omega_f}\right)_g = \tau_{S_{fg}/R_{fg}}^{\Omega_{fg}}$$

for all $f, g \in \mathfrak{F}$. Since the $D(f), f \in \mathfrak{F}$, are an open covering of $Spec(R)$ we deduce the existence of a unique R–linear map

$$\tau_{S/R}^{\Omega} : \Omega_S \to \Omega$$

such that $\left(\tau_{S/R}^{\Omega}\right)_f = \tau_{S_f/R_f}^{\Omega_f}$ for all $f \in \mathfrak{F}$.

It is clear from the construction of $\tau_{S/R}^{\Omega}$ that we may use $tr_{S/R}^k$ instead of $tr_{S/R}^{\mathbf{Z}}$, if Ω is a differential algebra of R/k.

4.1. Remark:

i) $\tau_{S/R}^{\Omega} : \Omega_S \to \Omega$ is homogeneous of degree 0.

ii) $\tau_{S/R}^{\Omega}$ is the pretrace of Angéniol ([A], (7.1.2); [AL], §6).

iii) Let R/k be an algebra and let S/R be finite and free. Let $\sigma : \Omega_{S/k}^{\cdot} \to \Omega_{R/k}^{\cdot}$ be a map, which makes the following diagram commutative:

$$
\begin{CD}
\Omega_{S/k}^{\cdot} @>{\theta_{S/k}^{\cdot}}>> H.(S/k) \\
@V{\sigma}VV @VV{tr_{S/R}}V \\
\Omega_{R/k}^{\cdot} @>{\theta_{R/k}^{\cdot}}>> H.(R/k)
\end{CD}
$$

i.e. σ is a ψ–trace in the sense of [L$_1$], (4.6.2), where ψ is a regular representation of S/R. Then:

$$
n!\sigma \left| \Omega_{S/k}^n = \tau_{S/R}^{\Omega_{R/k}^{\cdot}} \right| \Omega_{S/k}^n \text{ for all } n \in \mathbf{N}.
$$

i) and ii) are locally obvious, iii) is a consequence of ii) by [L$_1$], (4.6.5).

Next we will construct traces of differential forms for the classes of pairs $(S/R, \Omega)$ listed below. Following a suggestion of J. Lipman we will use traces in Hochschild homology for the construction. Most of the trace axioms then follow easily from the corresponding properties of traces in Hochschild homology.

Class I contains all pairs $(S/R, \Omega)$, where S/R is a finite projective algebra, Ω is a differential algebra of R and $\mathbf{Q} \subseteq R$. (B. Angéniol, [A], (7.1.2) and [AL], §6)

Class II contains all pairs $(S/R, \Omega)$, where R is a finite direct product of fields, S/R is a finite algebra, and Ω is a differential algebra of R. (E. Kunz, [KD], §16, ex. 5)

Class III contains all pairs $(S/R, \Omega)$, where S/R is finite and projective, 2 is not a zero–divisor on R or R is reduced and noetherian, and Ω is an exterior differential algebra of R such that Ω^1 is a finitely generated projective R–module. (J. Lipman, [L$_1$], (4.6.4); [AL], §6)

In §5 we will discuss how traces in Hochschild homology are related to traces of differential forms of complete intersections as constructed in [KD], §16.

4.2. Construction of traces for class I:

First let R be a $\mathbf{Q}$-algebra and let S/R be a finite free algebra. Then define

$$\sigma_{S/R}^{\Omega_{R/\mathbf{Z}}} : \Omega_{S/\mathbf{Z}}^{\cdot} \to \Omega_{R/\mathbf{Z}}^{\cdot}$$

by $\sigma_{S/R}^{\Omega_{R/\mathbf{Z}}} := \delta_{R/\mathbf{Z}} \circ tr_{S/R}^{\mathbf{Z}} \circ \theta_{S/\mathbf{Z}}^{\cdot}$ with the maps $\delta_{R/\mathbf{Z}}$ and $\theta_{S/\mathbf{Z}}^{\cdot}$ of (2.10) and (2.19).

If S/R is finite and projective we get by glueing a well–defined R–linear map

$$\sigma_{S/R}^{\Omega_{R/\mathbf{Z}}} : \Omega_{S/\mathbf{Z}}^{\cdot} \to \Omega_{R/\mathbf{Z}}^{\cdot}$$

by arguing as in the construction of the pretrace.

Now let Ω be an arbitrary differential algebra of R. Then $\Omega = \Omega_{R/\mathbf{Z}}^{\cdot}/\mathfrak{I}$, where $\mathfrak{I}$ is a homogeneous differentially closed ideal of $\Omega_{R/\mathbf{Z}}^{\cdot}$, and consequently $\Omega_S = \Omega_{S/\mathbf{Z}}^{\cdot}/\mathfrak{I}\Omega_{S/\mathbf{Z}}^{\cdot}$ ([KD], (3.4)). By (2.10), (2.17) and (3.4) (T1) $\sigma_{S/R}^{\Omega_{R/\mathbf{Z}}}\left(\mathfrak{I}\Omega_{S/\mathbf{Z}}^{\cdot}\right) \subseteq \mathfrak{I}$, and therefore we get a well–defined map

$$\sigma_{S/R}^{\Omega} : \Omega_S \to \Omega.$$

Clearly we may use $tr_{S/R}^{k}$ instead of $tr_{S/R}^{\mathbf{Z}}$ in the construction of $\sigma_{S/R}^{\Omega}$, if Ω is a differential algebra of R/k.

4.3. *Construction of traces for class II:*

If R is a finite direct product of fields, then

$$\theta_{R/\mathbf{Z}}^{\cdot} : \Omega_{R/\mathbf{Z}}^{\cdot} \to H.(R/\mathbf{Z})$$

is an isomorphism of DG–algebras by (2.23). So if S/R is a finite free algebra we can define

$$\sigma_{S/R}^{\Omega_{R/\mathbf{Z}}} : \Omega_{S/\mathbf{Z}}^{\cdot} \to \Omega_{R/\mathbf{Z}}^{\cdot}$$

by $\sigma_{S/R}^{\Omega_{R/\mathbf{Z}}} := \theta_{R/\mathbf{Z}}^{\cdot -1} \circ tr_{S/R}^{\mathbf{Z}} \circ \theta_{S/\mathbf{Z}}^{\cdot}$.

If S/R is an arbitrary finite algebra, then again we get by glueing a well defined map

$$\sigma_{S/R}^{\Omega_{R/\mathbf{Z}}} : \Omega_{S/\mathbf{Z}}^{\cdot} \to \Omega_{R/\mathbf{Z}}^{\cdot},$$

which as in (4.2) induces an R–homomorphism

$$\sigma_{S/R}^{\Omega} : \Omega_S \to \Omega$$

for an arbitrary differential–algebra Ω of R.

4.4. Construction of traces for class III:

First let S/R be finite and free, and let $\Omega^{\cdot}$ be an exterior differential algebra of R such that Ω^1 is a finite free R-module. Furthermore let $\mathfrak{A}$ be an R-basis of Ω^1, and let $\delta_{R/\mathbf{Z}}^{\cdot} : H.(R/\mathbf{Z}) \to \Omega^{\cdot}$ be the corresponding R-homomorphism of (2.21). Then there exists a map

$$f := \delta_{R/\mathbf{Z}}^{\cdot} \circ tr_{S/R}^{\mathbf{Z}} \circ \theta_{S/\mathbf{Z}}^{\cdot} : \Omega_{S/\mathbf{Z}}^{\cdot} \to \Omega.$$

If $\Omega = \Omega_{R/\mathbf{Z}}^{\cdot}/\mathfrak{I}$ with a differentially closed ideal $\mathfrak{I} \subseteq \Omega_{R/\mathbf{Z}}^{\cdot}$, then $\Omega_S = \Omega_{S/\mathbf{Z}}^{\cdot}/\mathfrak{I}\Omega_{S/\mathbf{Z}}^{\cdot}$, and (2.10), (2.21) and (3.4) (T1) imply: $f\left(\mathfrak{I}\Omega_{S/\mathbf{Z}}^{\cdot}\right) = 0$. Hence f induces a map

$$\sigma_{S/R}^{\Omega} : \Omega_S \to \Omega.$$

The corresponding properties of traces in Hochschild homology, (2.10) and (2.21) imply that this map satisfies the trace axioms TR 1, TR 5 and TR 7. However I don't know if $\sigma_{S/R}^{\Omega}$ in general is independent of the choice of a basis $\mathfrak{A}$ of Ω^1. If 2 is not a zero–divisor on R, or if R is reduced and noetherian, then $\delta_{R/\mathbf{Z}}^{\cdot}$ is independent of the basis of Ω^1 by (2.21) resp. (2.22), and so is $\sigma_{S/R}^{\Omega}$.

Let $(S/R, \Omega)$ be an arbitrary object of class III, and denote by $\mathfrak{F}$ the family of all $f \in R$ such that both S_f and Ω_{R_f} are free R-modules. By the above the maps $\sigma_{S_f/R_f}^{\Omega_f}$ glue to give

$$\sigma_{S/R}^{\Omega} : \Omega_S \to \Omega,$$

since $D(f), f \in \mathfrak{F}$, is an open covering of $Spec(R)$.

If Ω is a differential algebra of R/k, then we clearly may construct $\sigma_{S/R}^{\Omega}$ using

$$tr_{S/R}^{k} : H.(S/k) \to H.(R/k)$$

instead of $tr_{S/R}^{\mathbf{Z}}$.

Some properties of these trace morphisms follow immediately from the very definition:

4.5. Remark:

i) If $(S/R, \Omega)$ is object of two of the classes I–III, then the various definitions of $\sigma_{S/R}^{\Omega}$ coincide.

ii) If $(S/R, \Omega)$ is in one of the classes I–III, then it holds for every $n \in \mathbb{N}$:

$$n! \sigma_{S/R}^{\Omega} \Big| \Omega_S^n = \tau_{S/R}^{\Omega} \Big| \Omega_S^n.$$

iii) Traces are compatible with localization in the base ring. More precisely, if $(S/R, \Omega)$ is in one of the classes I–III, and if $N \subseteq R$ is a multiplicatively closed set, then $(S_N/R_N, \Omega_N)$ is in the same class, and the following diagram commutes:

$$
\begin{array}{ccc}
\Omega_S & \longrightarrow & \Omega_{S_N} \\
\Big\downarrow{\sigma_{S/R}^{\Omega}} & & \Big\downarrow{\sigma_{S_N/R_N}^{\Omega_N}} \\
\Omega & \longrightarrow & \Omega_N.
\end{array}
$$

Notation: If no confusion is likely we often will write $\sigma_{S/R}$ instead of $\sigma_{S/R}^{\Omega}$.

Of interest in connection with the theory developed by Lipman in [L₁] is the following simple statement:

4.6. PROPOSITION. *Let $(S/R, \Omega)$ be in class II. Furthermore suppose that R is a k-algebra and that $\Omega = \Omega_{R/k}^{\cdot}$. Then the following diagram commutes:*

$$
\begin{array}{ccc}
\Omega_{S/k}^{\cdot} & \xrightarrow{\;\;\theta_{S/k}^{\cdot}\;\;} & H.(S/k) \\
\Big\downarrow{\sigma_{S/R}} & & \Big\downarrow{tr_{S/R}} \\
\Omega_{R/k}^{\cdot} & \xrightarrow{\;\;\theta_{R/k}^{\cdot}\;\;} & H.(R/k).
\end{array}
$$

Proof: By (1.18) and [KD], (4.7) we may assume that R is a field. In this case the claim is evident, if k is the prime field of R, since then $\theta_{R/k}^{\cdot}$ is bijective. In general we may assume that $k \subseteq R$ is a subfield. Otherwise replace k by the quotient field of its image in R to get a diagram canonically isomorphic to the original one. Let $k_0 \subseteq k$ be the prime field of k. Then

we get a diagram

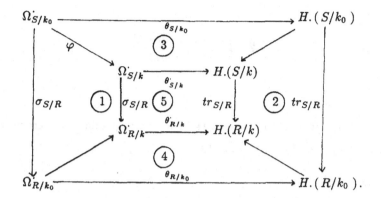

Subdiagram (1) commutes by construction, (2) by (3.4)(T3) and (3) and (4) by (2.10). Since φ is surjective amd since the exterior diagram commutes, diagram (5) also commutes.

A corresponding statement for traces for classes I resp. III is not evident and seemingly not known (see [L_1], (4.6)).

In the rest of this section we will try to prove the trace axioms for the traces constructed above.

4.7. Linearity.

If $(S/R, \Omega)$ is in one of the classes I–III, and if we consider Ω_S as a Ω–left module via the canonical map $\Omega \to \Omega_S$, then $\sigma^\Omega_{S/R}$ is Ω–linear and homogeneous of degree 0.

4.8. Relation to the canonical trace.

If $(S/R, \Omega)$ is in one of the classes I–III, then the restriction $\sigma^\Omega_{S/R}|\Omega^0_S : S \to R$ to the elements of degree 0 is the canonical trace $\sigma_{S/R}$.

4.9. Direct products.

Let $(S/R, \Omega)$ be in one of the classes I–III, and suppose $S = S_1 \times \cdots \times S_h$ is a direct product of R-algebras. Then $(S_1/R, \Omega), \ldots, (S_h/R, \Omega)$ are in the same class as $(S/R, \Omega)$, and the following diagram commutes:

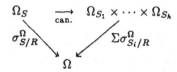

4.10. Logarithmic derivative.

If $(S/R, \Omega)$ is in one of the classes I–III, and if $a \in S$ is a unit, then it holds:

$$\sigma^{\Omega}_{S/R}\left(\frac{da}{a}\right) = \frac{dn_{S/R}(a)}{n_{S/R}(a)}.$$

(4.7)–(4.10) are immediate consequences of the corresponding properties of traces in Hochschild homology and of (2.10) and (2.19) resp. (2.21).

4.11. Base change.

Let $(S/R, \Omega)$ be in one of the classes I–III, let $\alpha : R \to R'$ be a homomorphism of rings, let $\beta : \Omega \to \Omega'$ be an α–homomorphism of Ω into a differential algebra Ω' of R', and set $S' := S \otimes_R R'$. If $(S'/R', \Omega')$ is in one of the classes I–III (not necessarily the same as $(S/R, \Omega)$), then the following diagram commutes:

$$
\begin{array}{ccc}
\Omega_S & \longrightarrow & \Omega'_{S'} \\
\downarrow{\scriptstyle \sigma^{\Omega}_{S/R}} & & \downarrow{\scriptstyle \sigma^{\Omega'}_{S'/R'}} \\
\Omega & \xrightarrow{\ \beta\ } & \Omega'
\end{array}
$$

Proof: By ((4.5)ii)) we may assume R and R' are local.

If both $(S'/R', \Omega')$ and $(S/R, \Omega)$ are in class II, then it suffices to prove the claim for $\Omega = \Omega_{R/\mathbb{Z}}$ and $\Omega' = \Omega_{R'/\mathbb{Z}}$. In this case it is an immediate consequence of (3.4) (T3), since $\theta_{R/\mathbb{Z}}$ and $\theta_{R'/\mathbb{Z}}$ are isomorphisms by (2.23).

If $(S/R, \Omega)$ is in class I, then necessarily $(S'/R', \Omega')$ is in class I. Again it suffices to show the claim for $\Omega = \Omega_{R/\mathbb{Z}}$ and $\Omega' = \Omega_{R'/\mathbb{Z}}$, in which case it follows easily from (2.10), (3.4) (T3) and (2.19).

If $(S/R, \Omega)$ is in II and if $(S'/R', \Omega')$ is in I, then $char(R) = 0$ and $(S/R, \Omega)$ is also in I. The claim follows from the above and from (4.5)i).

Let $(S/R, \Omega)$ be in II and $(S'/R', \Omega')$ in III. We may assume that $\Omega = \Omega_{R/\mathbb{Z}}$. Then we get a diagram:

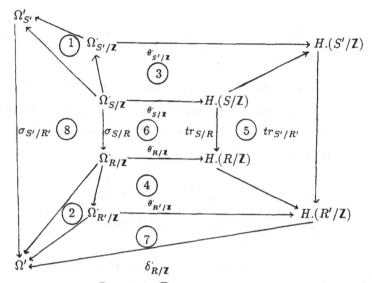

in which the subdiagrams ① and ② commute by [**KD**], (2.8)a). ③ and ④ commute by (2.10), ⑤ by (3.4) (T3), ⑥ by (4.6) and ⑦ by (2.21). Diagram chasing shows that also subdiagram ⑧ commutes.

Let $(S/R, \Omega)$ be in III and $(S'/R', \Omega')$ be in I. Then the morphisms $R \to R'$ resp. $\Omega \to \Omega'$ factor through $R \otimes_{\mathbf{Z}} \mathbf{Q}$ resp. $\Omega_{R \otimes_{\mathbf{Z}} \mathbf{Q}}$. Using (4.5)i) one sees that we therefore may assume: $\mathbf{Q} \subseteq R$. Then $(S/R, \Omega)$ is also in class I, and the claim follows from the above and (4.5)i).

Now let both $(S/R, \Omega)$ and $(S'/R', \Omega')$ be in class III, and suppose 2 is not a zero–divisor on R'. Let $\{\omega_1, \ldots, \omega_r\}$ be an R–basis of Ω^1, and let $\{d_1, \ldots, d_r\}$ be the corresponding system of derivations on R (i.e. $d_i = \omega_i^* \circ d (i = 1, \ldots, r)$, where $\{\omega_1^*, \ldots, \omega_r^*\}$ is the basis of $Hom_R(\Omega^1, R)$ dual to $\{\omega_1, \ldots, \omega_r\}$). Furthermore let $\{\eta_1, \ldots, \eta_s\}$ be an R'–basis of $(\Omega')^1$, and let $\{\delta_1, \ldots, \delta_s\}$ be the corresponding system of derivations on R'. The ring homomorphism $\alpha : R \to R'$ induces derivations $\overline{d_i} := \alpha \circ d_i (i = 1, \ldots, r)$ and $\overline{\delta_j} := \delta_j \circ \alpha (j = 1, \ldots, s)$ from R to R'.

In Ω' we have equations

$$\beta(\omega_i) = \sum_{j=1}^{s} r'_{i,j} \eta_j \text{ with } r'_{i,j} \in R' (i = 1, \ldots, r).$$

Therefore it holds in $Der(R, R')$:

$$\overline{\delta_j} = \sum_{i=1}^{r} r_{i,j} \overline{d_i} \qquad (j = 1, \ldots, s).$$

Let $\omega \in \Omega_S^p$ and denote by ω' its image in $\Omega_{S'}'$. Then it holds:

$$\sigma_{S'/R'}^{\Omega'}(\omega') = \left(\sum_{j_1 < \cdots < j_p} \delta_{j_1} \cdot \ldots \cdot \delta_{j_p} \otimes \eta_{j_1} \cdot \ldots \cdot \eta_{j_p} \right) \left(tr_{S'/R'}^{\mathbf{Z}} \left(\theta_{S'/\mathbf{Z}}'(\omega') \right) \right),$$

where a map is denoted by the same symbol as the element in the pth Hochschild cohomology group $H^p(R'/\mathbf{Z}, (\Omega')^p)$ which defines it via the pairing of ([$\mathbf{L_1}$], (1.1)). Since for every $\delta \in H^1(R'/\mathbf{Z}, R')$ it holds $2\delta^2 = 0$ in the ring $H^{\cdot}(R'/\mathbf{Z}, R')$, it follows easily from (2.10), (3.4) (T3) and the above relation, that

$$2\delta_{j_1} \cdot \ldots \cdot \delta_{j_p} \left(tr_{S'/R'}^{\mathbf{Z}} \left(\theta_{S'/\mathbf{Z}}'(\omega') \right) \right)$$
$$= \sum_{i_1 < \cdots < i_p} 2r_{i_1,\ldots,i_p}^{'j_1,\ldots,j_p} \beta \left(d_{i_1} \cdot \ldots \cdot d_{i_p} \left(tr_{S/R}^{\mathbf{Z}} \left(\theta_{S/\mathbf{Z}}'(\omega) \right) \right) \right),$$

where $r_{i_1,\ldots,i_p}^{'j_1,\ldots,j_p}$ is the determinant of the matrix obtained from the $i_1^{\text{st}},\ldots,i_p^{\text{th}}$ row and the $j_1^{\text{st}},\ldots,j_p^{\text{th}}$ column of the matrix $(r'_{i,j})$.

Furthermore it holds

$$\beta \left(\omega_{i_1} \cdot \ldots \cdot \omega_{i_p} \right) = \sum_{j_1 < \cdots < j_p} r_{i_1,\ldots,i_p}^{'j_1,\ldots,j_p} \eta_{j_1} \cdot \ldots \cdot \eta_{j_p} \qquad (i_1 < \cdots < i_p).$$

This implies that

$$\sigma_{S'/R'}^{\Omega'}(\omega') = \beta \left(\sigma_{S/R}^{\Omega}(\omega) \right)$$

since 2 is not a zero–divisor on R'.

Now let $(S/R, \Omega)$ be in class III and let $(S'/R', \Omega')$ be in class II. If $\mathfrak{P} := \alpha^{-1}(0) \in Spec(R)$, then it suffices (by (4.5)iii) to prove the claim for $(S_{\mathfrak{P}}/R_{\mathfrak{P}}, \Omega_{\mathfrak{P}})$ instead of $(S/R, \Omega)$. Hence we may assume: (R, m) is local and $m = \ker(\alpha)$.

Define $k := R/m$ and $\ell := S/mS$. Then $(\ell/k, \Omega)$ is in II, and therefore it suffices to show that the following diagram commutes:

$$\begin{array}{ccc} \Omega_S & \longrightarrow & \Omega_\ell \\ \downarrow{\scriptstyle \sigma_{S/R}^{\Omega}} & & {\scriptstyle \sigma_{\ell/k}^{\Omega_k}}\downarrow \\ \Omega & \longrightarrow & \Omega_k. \end{array}$$

From the fundamental exact sequence ([$\mathbf{KD}$], (4.17))

$$m/m^2 \xrightarrow{\alpha} \Omega^1/m\Omega^1 \xrightarrow{\bar{\beta}} \Omega_k^1 \to 0$$

we conclude that there exists a basis $\omega_1, \ldots, \omega_r$ of Ω^1 and an $s \leq r$ such that $\eta_1 := \bar{\beta}(\omega_1 + m\Omega^1), \ldots, \eta_s := \bar{\beta}(\omega_s + m\Omega^1)$ is a k–basis of Ω_k, and such that $\bar{\beta}(\omega_i + m\Omega^1) = 0$ for $i = s+1, \ldots, r$. Let $\{d_1, \ldots, d_r\}$ resp. $\{\delta_1, \ldots, \delta_s\}$ be the corresponding systems of derivations on R resp. k.

Using the basis $\eta_1, \ldots, \eta_s$ we can construct a map $\delta_{k/\mathbf{Z}} : H.(k/\mathbf{Z}) \to \Omega_k$ as in (2.21), and (2.21)i) implies that $\sigma_{\ell/k}^{\Omega_k}$ also can be described as $\delta_{k/\mathbf{Z}} \circ tr_{\ell/k}^{\mathbf{Z}} \circ \theta_{\ell/\mathbf{Z}}$. Using the notations of the previous proof we get in the equalities $\beta(\omega_i) = \sum_{i=1}^{s} r'_{i,j}\eta_j$ resp. $\bar{\delta}_j = \sum_{j=1}^{s} r'_{i,j}d_i$ for the coefficients: $r'_{i,j} = \delta_{i,j} (i = 1, \ldots, r; j = 1, \ldots, s)$.

Now it's an easy calculation to show that

$$
\delta_{j_1} \cdot \ldots \cdot \delta_{j_p} \left(tr_{\ell/k}^{\mathbf{Z}} \left(\theta_{\ell/\mathbf{Z}}(\omega') \right) \right)
$$
$$
= \sum_{i_1 < \cdots < i_p} r'^{j_1, \ldots, j_p}_{i_1, \ldots, i_p} \beta \left(d_{i_1} \cdot \ldots \cdot d_{i_p} \left(tr_{S/R}^{\mathbf{Z}} \left(\theta_{S/\mathbf{Z}}(\omega) \right) \right) \right)
$$

for $j_1 < \cdots < j_p$, and

$$
\beta \left(\omega_{i_1} \cdot \ldots \cdot \omega_{i_p} \right) = \sum_{j_1 < \cdots < j_p} r'^{j_1, \ldots, j_p}_{i_1, \ldots, i_p} \eta_{j_1} \cdot \ldots \cdot \eta_{j_p}
$$

showing that

$$
\sigma_{\ell/k}^{\Omega_k}(\omega') = \beta \left(\sigma_{S/R}^{\Omega}(\omega) \right).
$$

Finally let $(S/R, \Omega)$ and $(S'/R', \Omega')$ be in class III, and suppose that R' is reduced and noetherian. Denote by $K := Q(R)$ the full ring of fractions, and by $L := S' \otimes_{R'} K$.

Since $\Omega' \to \Omega'_K$ is injective, it suffices (by (4.5)iii)) to show the claim for $(L/K, \Omega'_K)$ instead of $(S'/R', \Omega')$. Since $(L/K, \Omega'_K)$ is in class II, the assertion follows from (4.5)i) and the above.

4.12. Local algebras

Let $(S/R, \Omega)$ be in one of the classes I–III. Suppose that R and S are local with maximal ideals m_R and m_S, let $K := R/m_R, L := S/m_S$ and $n := \operatorname{rank}_R(S) \cdot \dim_K(L)^{-1}$. Then $(L/K, \Omega_K)$ is in class II, and the following diagram commutes:

$$
\begin{array}{ccc}
\Omega_S & \longrightarrow & \Omega_L \\
\sigma_{S/R}^{\Omega} \downarrow & & \downarrow n\sigma_{L/K}^{\Omega_k} \\
\Omega & \longrightarrow & \Omega_K.
\end{array}
$$

Proof: By (4.11) we may assume that $R = K$. In this case $(S/R, \Omega)$ also is in class II, and it suffices to show the claim for $\Omega = \Omega'_{R/\mathbf{Z}}$. Since $\theta'_{R/\mathbf{Z}}$ is an isomorphism by (2.23), it now is an immediate consequence of (3.4) (T8).

4.13. Transitivity

Suppose that one of the following conditions holds:

i) $(S/R, \Omega)$ is in one of the classes I–III, $(T/S, \Omega_S)$ is in class II.

ii) $(S/R, \Omega)$ is in class II, $(T/S, \Omega_S)$ is in one of the classes I–III.

iii) $(S/R, \Omega)$ is in class I, R is reduced and noetherian and Ω is torsion free, and $(T/S, \Omega_S)$ is in one of the classes I–III.

iv) $(S/R, \Omega)$ is in class III, R is reduced and noetherian, and $(T/S, \Omega_S)$ is in one of the classes I–III.

Then it holds:

$$\sigma^{\Omega}_{T/R} = \sigma^{\Omega}_{S/R} \circ \sigma^{\Omega_S}_{T/S}.$$

Proof: We always may assume: R is local.

i) In this case R necessarily is a field, and therefore $(S/R, \Omega)$ is in class II. It suffices to show the claim for $\Omega = \Omega'_{R/\mathbf{Z}}$, and furthermore we may assume that S is a field (TR3). Now it follows from (3.4)(T5) and (4.6).

ii) In this case S and T are artinian rings. Using (4.11), (4.9) and [KD], (4.7) one easily reduces to the case that S and T are local. Let m_S resp. m_T be the maximal ideals of S resp. $T, L := S/m_S, M := T/m_S T, \overline{M} := T/m_T, n := \dim_R(S) \cdot \dim_R(L)^{-1}$ and $m := \dim_L(M) \cdot \dim_L(\overline{M})^{-1}$. Then $(L/R, \Omega), (M/L, \Omega_L)$ and $(\overline{M}/L, \Omega_L)$ are in class II, and by (4.12) and (4.11) the following diagrams commute:

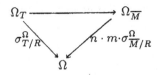

and

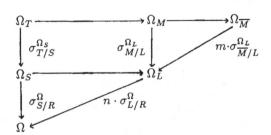

Therefore we may assume that S and T are fields, so that condition i) is satisfied. The assertion follows from the previous proof.

Suppose condition iii) or iv) is satisfied, and denote by $K = Q(R)$ the full ring of fractions of R. Then $\Omega \to \Omega_K$ is injective, and therefore we may replace $(S/R, \Omega)$ resp. $(T/S, \Omega_S)$ by $(S \otimes_R K/K, \Omega_K)$ resp. $(T \otimes_R K/S \otimes_R K, \Omega_{S \otimes K})$. K is a direct product of fields, and the assertion follows, since it holds if condition ii) is satisfied.

4.14. Compatibility with differentiation

Suppose that one of the following conditions hold:

i) $(S/R, \Omega)$ is in class I or II.

ii) $(S/R, \Omega)$ is in class III and R is reduced and noetherian.

Let d be the differentiation of Ω, and let d_S be the differentiation of Ω_S. Then

$$d \circ \sigma^{\Omega}_{S/R} = \sigma^{\Omega}_{S/R} \circ d_S.$$

Proof: We always may assume: R is local.

i) We may in addition assume: $\Omega = \Omega'_{R/\mathbf{Z}}$. Then the claim follows easily from the corresponding statement for traces in Hochschild homology and from (2.10) and (2.13) resp. (2.19).

ii) As in the proof of (4.13) we may replace R by its full ring of fractions, so that condition i) is satisfied.

The trace axioms TR1–TR8 are natural properties for traces of differential forms. Therefore it is a natural question, to ask whether the trace already is determined by these properties. I do not know the answer to this question, however partial results are obtained below.

Let X be a class whose objects $(S/R, \Omega)$ have the following properties:

E1) R is reduced and noetherian, Ω is a differential algebra of R and torsion free as an R-module, and S/R is a finite projective algebra.

E2) If $N \subseteq R$ is a multiplicatively closed set, then $(S_N/R_N, \Omega_N)$ also is in class X.

E3) If $S = S_1 \times \cdots \times S_h$ is a finite direct product of R–algebras, then $(S_i/R, \Omega)$ is in X for every $i \in \{1, \ldots, h\}$.

E4) If R is a field and if S is local with maximal ideal m, then $(S/m/R, \Omega)$ also is in X.

E5) If R and S are fields, and if T is an intermediate field of S/R, then both $(T/R, \Omega)$ and $(S/T, \Omega_T)$ are in X.

4.15. THEOREM. *Let X be a class of objects $(S/R, \Omega)$ satisfying E1)–E5). Then there exists at most one system of trace maps*

$$\sigma^{\Omega}_{S/R} : \Omega_S \to \Omega, \quad (S/R, \Omega) \text{ in } X,$$

satisfying TR1–TR8.

Proof: Let $\left\{ \sigma^{\Omega}_{S/R} : (S/R, \Omega) \text{ in } X \right\}$ and $\left\{ \tilde{\sigma}^{\Omega}_{S/R} : (S/R, \Omega) \text{ in } X \right\}$ be two systems of trace maps satisfying TR1–TR8. Let $(S/R, \Omega)$ be an object of X, and denote by K the full ring of fractions of R. Then $\Omega \to \Omega_K$ is injective by E1). In order to show $\sigma^{\Omega}_{S/R} = \tilde{\sigma}^{\Omega}_{S/R}$ we therefore may replace R by K (by E2) and TR3) and hence assume that R is a finite direct product of fields. Again by E2) and TR3 it suffices to show the equality locally, i.e. we may assume R is a field. Then S is a finite direct product of local R–algebras. Since both systems of traces satisfy TR4 we may assume by E3): S is local. Using TR8 and E4) we may replace S by its residue class field, and it remains to show:

$$\sigma^{\Omega}_{S/R} = \tilde{\sigma}^{\Omega}_{S/R} \text{ for all those } (S/R, \Omega) \text{ in } X \text{ for which } S/R \text{ is a finite field extension.}$$

For such an $(S/R, \Omega)$ we have by E5) for any intermediate field T of S/R: Both $(T/R, \Omega)$ and $(S/T, \Omega_T)$ are also in X. Therefore it suffices by TR5 to show the equality in the following two cases:

a) S/R is a separable field extension.

b) S/R is purely inseparable of degree p, where $p = \operatorname{char}(R)$.

In these cases it follows from TR1, TR2, TR6 and TR7 ([K_4], (2.3.6)).

Now let I' be the class of all $(S/R, \Omega)$ in I, for which R is reduced and noetherian and Ω is torsion free. Furthermore let III' be the class of all $(S/R, \Omega)$ in III, for which R is reduced and noetherian. Each of the classes I', II and III' satisfies E1)–E5), and therefore it follows from (4.7)–(4.15):

4.16. COROLLARY. *For each of the classes I', II, III' there exists precisely one system of trace maps $\left\{ \sigma^{\Omega}_{S/R} \right\}$ satisfying TR1–TR8.*

4.17. Remark:

i) A general proof of transitivity of traces resp. of their compatibility with differentiation is not known to me. Also the problem of uniqueness could not be solved completely.

ii) In §5 transitivity of traces resp. their compatibility with differentiation will be proved in some additional situations. For instance if $(S/R, \Omega)$ is in class III, if R is noetherian without embedded primes, and if S/R is generically a complete intersection, then $\sigma^\Omega_{S/R}$ commutes with differentiation.

iii) In §5 it also will be shown that the condition "R is reduced and noetherian or 2 is not a zero–divisor on R" for objects $(S/R, \Omega)$ of class III can be weakened slightly.

iv) For every ring R the map

$$\theta^1_{R/\mathbf{Z}} : \Omega^1_{R/\mathbf{Z}} \to H_1(R/\mathbf{Z})$$

is bijective by (2.18). Therefore we always can construct traces of differential forms of degree 0 and 1. These maps satisfy TR1–TR4, TR6–TR8 and under weak assumptions also TR5.

§5. Traces in complete intersections.

In [KD], §16. E. Kunz constructs a system of traces

$$\sigma^{\Omega}_{S/R} : \Omega_S \to \Omega$$

for the class IV of all pairs $(S/R, \Omega)$ where

a) R is a noetherian ring,

b) S/R is finite and locally a complete intersection ([KD], (C.3)),

c) Ω is an arbitrary differential algebra of R.

He also shows that this system satisfies the axioms TR1–TR7 and is uniquely determined by TR1–TR4.

The relations between these traces and traces in Hochschild homology will be established in the following theorem. Its proof heavily relies on the theory of Hochschild homology and differential algebras of topological algebras as developed in §§1–3.

5.1. THEOREM. *Let R/k be an algebra. Suppose R is local and noetherian and S/R is a finite complete intersection. Then the following diagram commutes:*

$$
\begin{array}{ccc}
\Omega'_{S/k} & \xrightarrow{\;\;\theta'_{S/k}\;\;} & H.(S/k) \\[4pt]
{\scriptstyle \sigma^{\Omega_{R/k}}_{S/R}}\downarrow & {\scriptstyle tr^k_{S/R}} & \downarrow \\[4pt]
\Omega'_{R/k} & \xrightarrow{\;\;\theta'_{R/k}\;\;} & H.(R/k),
\end{array}
$$

i.e. $\sigma^{\Omega_{R/k}}_{S/R}$ is a ψ–trace in the sense of [L], (4.6.2), where ψ is a regular representation of S/R.

Proof: In the diagram

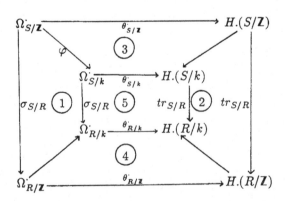

the subdiagram ① commutes by TR3, ② commutes by (3.4) (T3), and ③ and ④ commute by (2.10). (As in §§3 and 4 we omit superscripts if no confusion is likely.) In order to show that ⑤ commutes we therefore may assume that $k = \mathbf{Z}$, since φ is surjective.

It suffices to prove the claim element by element. Suppose $\omega \in \Omega_{S/\mathbf{Z}}^{\cdot}$ is given. Then it holds:

5.2. LEMMA. *There exists a local ring R' that is essentially of finite type over $\mathbf{Z}$, and there exists a finite algebra S'/R' such that:*

i) *S'/R' is a complete intersection.*

ii) *There exists a local homomorphism $R' \to R$ such that $S = S' \otimes_R R'$.*

iii) *ω is in the image of $\Omega_{S'/\mathbf{Z}}^{\cdot} \to \Omega_{S/\mathbf{Z}}^{\cdot}$.*

Proof: [**KD**], (16.3)

Let S'/R' be as in the lemma. Then we get a diagram

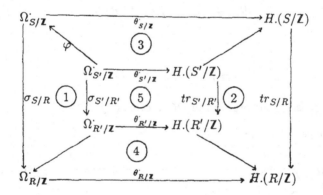

in which subdiagram ① commutes by TR3 (base change), ② commutes by (3.4)(T3) and ③ and ④ commute by (2.10). In order to show that

$$tr_{S/R}\left(\theta_{S/\mathbf{Z}}^{\cdot}(\omega)\right) = \theta_{R/\mathbf{Z}}^{\cdot}\left(\sigma_{S/R}^{\Omega_{R/\mathbf{Z}}^{\cdot}}(\omega)\right)$$

it suffices to show that subdiagram ⑤ commutes since $\omega \in \operatorname{im}(\varphi)$. Therefore we may assume that $R/\mathbf{Z}$ is essentially of finite type over $\mathbf{Z}$.

Next we need the following lemma, which is a modification of [**KD**], (16.2):

5.3. LEMMA. *In the above situation there exists a regular local ring R_1 with char $(R_1) = 0$ which is essentially of finite type and smooth over $\mathbf{Z}$, a surjective homomorphism $\varphi : R_1 \to R$ of local rings and an algebra S_1/R_1 with S_1 being essentially of finite type over $\mathbf{Z}$ and regular, satisfying:*

If $\mathfrak{J} = \ker(\varphi)$, and if R_0 is the $\mathfrak{J}$-adic completion of R_1 and S_0 is the $\mathfrak{J}S_1$-adic completion of S_1, then it holds:

i) *S_0/R_0 is a finite complete intersection.*

ii) *$S = R \otimes_{R_0} S_0$.*

iii) *R_0 is a regular local ring with char $(R_0) = 0$ and S_0 is a regular semilocal ring.*

The proof of this lemma will be given after the proof of (5.1).

In the situation of (5.3) the universally finite differential algebras $\tilde{\Omega}_{R_0/\mathbf{Z}}$ and $\tilde{\Omega}_{S_0/\mathbf{Z}}$ exist and $\tilde{\Omega}_{R_0/\mathbf{Z}}$ is the completion of $\Omega_{R_1/\mathbf{Z}}$, hence free as an R_0-module ([**KD**], (12.4)). If τ denotes the $\mathfrak{J}R_0$-adic topology on R_0 and τ' denotes the $\mathfrak{J}S_0$-adic topology on S_0, then $(R_0, \tau)/\mathbf{Z}$ and $(S_0, \tau')/\mathbf{Z}$ satisfy the conditions of (2.9), and therefore the morphisms

$$\theta_{(R_0/\mathbf{Z},\tau)}^{\cdot} : \tilde{\Omega}_{R_0/\mathbf{Z}}^{\cdot} \to H_{\cdot}(R_0/\mathbf{Z}, \tau)$$

and

$$\theta_{(S_0/\mathbf{Z},\tau')}^{\cdot} : \tilde{\Omega}_{S_0/\mathbf{Z}}^{\cdot} \to H_{\cdot}(S_0/\mathbf{Z}, \tau')$$

exist, and $\theta_{(R_0/\mathbf{Z},\tau)}^{\cdot}$ is an isomorphism of DG-algebras by (2.14).

Therefore we get a diagram

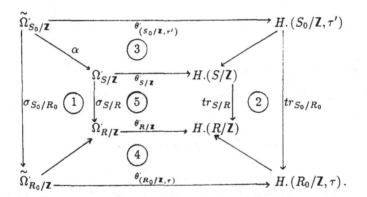

As above the subdiagrams ①, ②, ③ and ④ commute. In order to show that ⑤ commutes, it suffices to show that the outer square commutes, since α is surjective.

Let $N := R_0 \backslash \{0\}$, $K := (R_0)_N = Q(R_0)$ and $L := S_0 \otimes_{R_0} K$. Since $\tilde{\Omega}_{R_0/\mathbb{Z}}$ is free as an R_0-module, and since $\theta_{(R_0/\mathbb{Z},\tau)}$ is an isomorphism, the canonical map

$$H.(R_0/\mathbb{Z}, \tau) \to H.(R_0/\mathbb{Z}, \tau)_N$$

is injective. Therefore it suffices to show that the following diagram commutes:

$$
\begin{array}{ccc}
\left(\tilde{\Omega}_{S_0/\mathbb{Z}} \right)_N & \xrightarrow{\left(\theta_{(S_0/\mathbb{Z},\tau')} \right)_N} & H.(S_0/\mathbb{Z}, \tau')_N \\
\downarrow {(\sigma_{S_0/R_0})_N} & \qquad {(tr_{S_0/R_0})_N} & \downarrow \\
\left(\tilde{\Omega}_{R_0/\mathbb{Z}} \right)_N & \xrightarrow{\left(\theta_{(R_0/\mathbb{Z},\tau)} \right)_N} & H.(R_0/\mathbb{Z}, \tau)_N .
\end{array}
$$

K is a field of characteristic 0, since R_0 is a regular local ring with char $(R_0) = 0$, and $L = Q(S_0) = L_1 \times \cdots \times L_t$ ([**KD**], (A.6)) with separable algebraic field extensions L_i/K, since being a semilocal regular ring S_0 is reduced. In particular L/K is étale.

Since $\tilde{\Omega}_{S_0/\mathbb{Z}}$ is the universal S_0–extension of $\tilde{\Omega}_{R_0/\mathbb{Z}}$ by [**KD**], (11.9) it follows from [**KD**], (4.22) and [**KD**], (6.16):

$$\left(\tilde{\Omega}_{S_0/\mathbb{Z}} \right)_N \cong L \otimes_K \left(\tilde{\Omega}_{R_0/\mathbb{Z}} \right)_N .$$

All maps in the above diagram are homogeneous of degree 0 and $\left(\tilde{\Omega}_{R_0/\mathbb{Z}} \right)_N$ –linear and satisfy

$$(tr_{S_0/R_0})_N \Big|_{H_0(S/\mathbb{Z},\tau')_N} = \sigma_{L/K} = \left(\sigma_{S_0/R_0}^{\tilde{\Omega}_{R_0/\mathbb{Z}}} \right)_N \Big| \left(\tilde{\Omega}_{S_0/\mathbb{Z}}^0 \right)_N .$$

Since $\theta_{(S_0/\mathbb{Z},\tau')}^0 = id_{S_0}$ and $\theta_{(R_0/\mathbb{Z},\tau)}^0 = id_{R_0}$, the above diagram commutes in degree 0. Since $\left(\tilde{\Omega}_{S_0/\mathbb{Z}} \right)_N \cong L \otimes_K \left(\tilde{\Omega}_{R_0/\mathbb{Z}} \right)_N$ this implies the commutativity of the whole diagram, completing the proof of theorem 5.1.

Proof of (5.3): Write $R = \mathbb{Z}[X_1, \ldots, X_\ell]_{\mathfrak{p}} / \mathfrak{A}$ for some $\mathfrak{p} \in \text{Spec}(\mathbb{Z}[X_1, \ldots, X_\ell])$ and some ideal $\mathfrak{A}$ of $\mathbb{Z}[X_1, \ldots, X_\ell]_{\mathfrak{p}}$. By [**KD**], (C.14) the algebra S/R has a presentation

$$S = R[Y_1, \ldots, Y_n] / (t_1, \ldots, t_n)$$

as a complete intersection. Write

$$t_i = \sum_{|\nu| \leq N} \rho_\nu^{(i)} Y_1^{\nu_1} \cdot \ldots \cdot Y_n^{\nu_n}, \rho_\nu^{(i)} \in R, i = 1, \ldots, n.$$

Define $R_3 := \mathbf{Z}[X_1, \ldots, X_\ell]_\mathfrak{p}$ and $R_2 := R_3 \left[\left\{ Y_\nu^{(i)} \right\}_{|\nu| \leq N, i=1, \ldots, n} \right]$. Then there exists an epimorphism $\varphi : R_2 \to R$ extending the canonical epimorphism $R_3 \to R$ and satisfying $\varphi \left(Y_\nu^{(i)} \right) = \rho_\nu^{(i)}$. Let

$$T_i := \sum_{|\nu| \leq N} Y_\nu^{(i)} Z_1^{\nu_1} \cdot \ldots \cdot Z_n^{\nu_n} \quad (i = 1, \ldots, n),$$

and let

$$S_2 := R_2[Z_1, \ldots, Z_n] / (T_1, \ldots, T_n).$$

Then $S_2 \cong R_3 \left[\left\{ Y_\nu^{(i)} \right\}_{\nu \neq 0}, Z_1, \ldots, Z_n \right]$. In particular R_2 and S_2 are regular domains, and both are smooth over $\mathbf{Z}$.

Let $\mathfrak{m}$ be the maximal ideal of R, and let $\mathfrak{M} := \varphi^{-1}(\mathfrak{m})$. It follows immediately from the construction of φ that $\mathfrak{M} \in \mathrm{Max}\,(R_2)$, and that $\mathfrak{M} \cap R_3 = \mathfrak{p}R_3$. In particular this implies: $\dim((R_2)_\mathfrak{M}) = \dim(R_3) + $ number of unknowns $Y_\nu^{(i)}$.

Define $R_1 := (R_2)_\mathfrak{M}, r := \dim((R_2)_\mathfrak{M}), S_1 := (S_2)_\mathfrak{M}$ and $\mathfrak{I} := \ker(\varphi_\mathfrak{M} : R_1 \to R)$. Then it holds:

$\alpha)$ R_1 is a regular local ring, essentially of finite type over $\mathbf{Z}$ and smooth over $\mathbf{Z}$ at $\mathfrak{M}R_1$.

$\beta)$ S_1 is a regular domain and it is essentially of finite type over $\mathbf{Z}$.

$\gamma)$ $R = R_1/\mathfrak{I}$ and $S = R \otimes_{R_1} S_1 = S_1/\mathfrak{I}S_1$.

Denote by R_0 the $\mathfrak{I}$–adic completion of R_1 and by S_0 the $\mathfrak{I}S_1$–adic completion of S_1. Then $\alpha), \beta)$ and $\gamma)$ imply:

a) $R_0/\mathfrak{I}R_0 = R_1/\mathfrak{I} = R$ and $S_0 \otimes_{R_0} R = S_0/\mathfrak{I}S_0 = S$ by ([**GS**], (2.15) and (4.3)).

b) S_0/R_0 is finite by [**KD**], (12.8), since S/R is.

c) R_0 is a regular local ring of characteristic 0 by [**GS**], (8.3).

d) S_0 is a semilocal regular ring, and for every $\mathfrak{N} \in \mathrm{Max}\,(S_0)$ it holds: $\dim((S_0)_\mathfrak{N}) = \dim(S_0) = \dim(R_0) = r$:

S_0 is regular by [**GS**], (8.3), and by [**GS**], (2.19) we get canonical bijections $\mathrm{Max}\,(S_0) \cong \mathrm{Max}\,(S_1) \cap \mathfrak{V}(\mathfrak{I}S_1) \cong \mathrm{Max}\,(S_1/\mathfrak{I}S_1) \cong \mathrm{Max}\,(S)$, implying that S_0 is semilocal.

Therefore we get for $\mathfrak{N} \in \mathrm{Max}\,(S_0)$ that $\mathfrak{n} := \mathfrak{N} \cap S_1 \in \mathrm{Max}\,(S_1)$, and that $\mathfrak{n} \supseteq \mathfrak{I}S_1$. Furthermore the image $\bar{\mathfrak{n}}$ of $\mathfrak{n}$ in S is a maximal ideal, hence there exists $\tilde{\mathfrak{n}} \in \mathrm{Max}\,(S_2)$, $\tilde{\mathfrak{n}} \supseteq \ker(S_2 \to S)$ such that $\mathfrak{n} = \tilde{\mathfrak{n}}S_1$. Since $\bar{\mathfrak{n}} \cap R = \mathfrak{m}$ we get $\tilde{\mathfrak{n}} \cap R_2 = \mathfrak{M}$, and therefore $\tilde{\mathfrak{n}} \cap R_3 = \mathfrak{p}R_3$. Then it is well known that $\dim\,((S_1)_{\mathfrak{n}}) = \dim\left((S_2)_{\tilde{\mathfrak{n}}}\right) = r$, and therefore $\dim\,((S_0)_{\mathfrak{N}}) = r$ by ([**GS**] (6.2)c) and (7.3)). Similarly one shows that $\dim\,(R_0) = r$.

e) S_0/R_0 is flat:

For this let $\mathfrak{N} = \mathfrak{M}R_0$ be the maximal ideal of R_0, let $\hat{R}_0$ be the $\mathfrak{N}$–adic completion of R_0, and let $\widehat{S_0}$ be the $\mathfrak{N}S_0$–adic completion of S_0. Then $\widehat{S_0}$ is a finite direct product of regular local rings, all of them having the same dimension as $\widehat{R}_0$ (by d) and [**GS**], (6.2), (7.3)). A formula of Auslander–Buchsbaum ([**K$_1$**], VII.(1.12)) now implies that $\widehat{S_0}$ is a free $\widehat{R}_0$–module. Since $\widehat{S_0} = S_0 \otimes_{R_0} \widehat{R}_0$, it follows by [**M**], (4.E) that S_0/R_0 is flat.

f) S_0/R_0 is a finite complete intersection:

By [**KD**], (C.4) and e) S_0/R_0 is locally a complete intersection, since $S_0/\mathfrak{N}S_0$ is a complete intersection over $R_0/\mathfrak{M}R_0$. Therefore S_0/R_0 is a complete intersection by [**KD**], (C.13) and d).

This completes the proof of (5.3).

5.4. **Remark:** Let R be a noetherian k–algebra, and suppose that either R is a finite direct product of local k–algebras or that R is flat over k. If S/R is a finite locally complete intersection, then the following diagram commutes:

$$
\begin{array}{ccc}
\Omega^{\cdot}_{S/k} & \xrightarrow{\ \theta^{\cdot}_{S/k}\ } & H.(S/k) \\[2mm]
{\scriptstyle \sigma^{\cdot}_{S/R}}\Big\downarrow{\scriptstyle \Omega^{\cdot}_{R/k}} & {\scriptstyle tr^{k}_{S/R}} & \Big\downarrow \\[2mm]
\Omega^{\cdot}_{R/k} & \xrightarrow{\ \theta^{\cdot}_{R/k}\ } & H.(R/k).
\end{array}
$$

Proof: In both cases TR3 (base change) and (3.4)(T3) imply that it suffices to prove the claim locally in R (using (1.18) and [**KD**], (4.7) resp. (1.15)). Then the assertion follows from (5.1).

5.5. **COROLLARY.** *Under the assumptions of (5.1) suppose that $\mathfrak{I} \subseteq R$ is an ideal of R, and that τ is the $\mathfrak{I}$–adic topology on R. If the assumptions of (2.9) are satisfied for $(R,\tau)/k$,* and if τ' is the $\Im S$–adic topology on S, then the assumptions of (2.9) are also satisfied for $(S, \tau')/k$, and the following diagram commutes:

$$
\begin{array}{ccc}
\widetilde{\Omega}^{\cdot}_{\hat{S}/k} & \xrightarrow{\quad\theta_{(S/k,\tau')}\quad} & H.(S/k, \tau') \\
\Big\downarrow{\scriptstyle\sigma^{\widetilde{\Omega}_{R/k}}_{\hat{S}/\hat{R}}} & {\scriptstyle tr^k_{(S/R,\tau)}}\Big\downarrow & \\
\widetilde{\Omega}^{\cdot}_{\hat{R}/k} & \xrightarrow{\quad\theta_{(R/k,\tau)}\quad} & H.(R/k, \tau).
\end{array}
$$

Proof: We may assume that R and S are complete in their topologies. Then we get a diagram

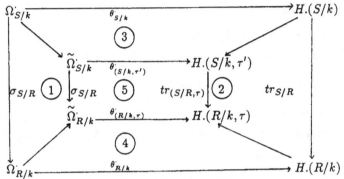

in which the subdiagrams ①, ②, ③ and ④ commute by TR3, (3.4) (T3) and (2.10). Since α is surjective, the commutativity of ⑤ follows from (5.1).

5.6. COROLLARY. *Let R be noetherian, let S/R be a finite locally complete intersection, and let Ω be a differential algebra of R.*
i) *If $\mathbb{Q} \subseteq R$, then the trace $\sigma^{\Omega}_{S/R} : \Omega_S \to \Omega$ of [KD], §16 coincides with the trace of (4.2).*
ii) *If $(S/R, \Omega)$ is in class II or III, then the trace $\sigma^{\Omega}_{S/R} : \Omega_S \to \Omega$ of [KD], §16 coincides with the trace of (4.3) resp. (4.4).*

Proof: The claims may be checked locally in R, in which case they follow easily from (5.1) and the constructions in (4.2), (4.3) and (4.4).

5.7. Remark: Let S/R be a finite complete intersection, and suppose that R and S are local and noetherian with maximal ideals $\mathfrak{m}_R$ and $\mathfrak{m}_S$. Let $k := R/\mathfrak{m}_R, \ell := S/\mathfrak{m}_S$ and $n := \operatorname{rank}_R(S) \cdot \dim_k(\ell)^{-1}$. Let Ω be a differential algebra of R. Then the following diagram commutes:

$$
\begin{array}{ccc}
\Omega_S & \longrightarrow & \Omega_\ell \\
\Big\downarrow{\scriptstyle\sigma^{\Omega}_{S/R}} & {\scriptstyle n\sigma^{\Omega_k}_{\ell/k}}\Big\downarrow & \\
\Omega & \longrightarrow & \Omega_k.
\end{array}
$$

Proof: By TR3 we may assume that $R = k$. In this case $(S/R, \Omega)$ is in class II, and the claim follows from (4.12) by (5.6).

Using (5.1) it is possible to deduce certain properties of traces constructed via Hochschild homology from the corresponding properties of traces in complete intersections, provided that the algebras in question are generically complete intersections.

5.8. PROPOSITION. *Let R be noetherian with $Ass(R) = Min(R)$, and let S/R be finite, free and generically a complete intersection. Suppose that $\Omega = \Lambda_R^{\cdot}(\Omega^1)$ is an exterior differential algebra of R with Ω^1 being finite and free as an R-module. Let $\mathfrak{A}$ be a basis of Ω^1 and let $\delta_{R/\mathbf{Z}}^{\cdot} : H_{\cdot}(R/\mathbf{Z}) \to \Omega$ be the map of (2.21), constructed using $\mathfrak{A}$. Then the composition $\delta_{R/\mathbf{Z}}^{\cdot} \circ tr_{S/R}^{\mathbf{Z}} \circ \theta_{S/\mathbf{Z}}^{\cdot} : \Omega_{S/\mathbf{Z}}^{\cdot} \to \Omega$ is independent of the basis $\mathfrak{A}$ of Ω^1 and induces a well defined map*

$$\sigma_{S/R}^{\Omega} : \Omega_S \to \Omega$$

Denoting by $K := Q(R)$ the full ring of fractions of R and by $L := Q(R) \otimes_R S = Q(S)$, then the following diagram commutes:

$$
\begin{array}{ccc}
\Omega_S & \longrightarrow & \Omega_L \\
\downarrow{\scriptstyle \sigma_{S/R}^{\Omega}} & {\scriptstyle \sigma_{L/K}^{\Omega_K}}\downarrow & \\
\Omega & \longrightarrow & \Omega_K,
\end{array}
$$

with $\sigma_{L/K}^{\Omega_K}$ being the trace of [KD], §16.

Proof: The canonical morphism $\Omega \to \Omega_K$ is injective and $\mathfrak{A}$ induces a K-basis of Ω_K^1 which again will be denoted by $\mathfrak{A}$. If $\delta_{K/\mathbf{Z}}^{\cdot} : H_{\cdot}(K/\mathbf{Z}) \to \Omega_K$ is the map of (2.21), constructed using this basis $\mathfrak{A}$, then the following diagram commutes:

$$
\begin{array}{ccc}
H_{\cdot}(R/\mathbf{Z}) & \xrightarrow{\ \delta_{R/\mathbf{Z}}^{\cdot}\ } & \Omega \\
\downarrow{\scriptstyle \text{can.}} & & \downarrow{\scriptstyle \text{can.}} \\
H_{\cdot}(K/\mathbf{Z}) & \xrightarrow{\ \delta_{K/\mathbf{Z}}^{\cdot}\ } & \Omega_K
\end{array}
$$

Therefore it suffices to show that the map

$$\delta_{K/\mathbf{Z}}^{\cdot} \circ tr_{L/K}^{\mathbf{Z}} \circ \theta_{L/\mathbf{Z}}^{\cdot} : \Omega_{L/\mathbf{Z}}^{\cdot} \to \Omega_K$$

is independent of the basis of Ω_K^1, used in its construction, and that it induces the trace

$$\sigma_{L/K}^{\Omega_K} : \Omega_L \to \Omega$$

of [KD], §16.

By (2.21) the following diagram commutes

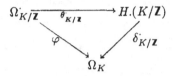

where φ is the canonical map.

By (5.4) also the following diagram commutes:

$$\begin{array}{ccc}
\Omega_{L/\mathbf{Z}}^{\cdot} & \xrightarrow{\;\;\theta_{L/\mathbf{Z}}^{\cdot}\;\;} & H_{\cdot}(L/\mathbf{Z}) \\
\downarrow{\sigma_{L/K}} & & \downarrow{tr_{L/K}} \\
\Omega_{K/\mathbf{Z}}^{\cdot} & \xrightarrow{\;\;\theta_{K/\mathbf{Z}}^{\cdot}\;\;} & H_{\cdot}(K/\mathbf{Z}).
\end{array}$$

Therefore it holds:

$$\delta_{K/\mathbf{Z}}^{\cdot} \circ tr_{L/K}^{\mathbf{Z}} \circ \theta_{L/\mathbf{Z}}^{\cdot} = \delta_{K/\mathbf{Z}}^{\cdot} \circ \theta_{K/\mathbf{Z}}^{\cdot} \circ \sigma_{L/K}^{\Omega_{K/\mathbf{Z}}^{\cdot}} = \varphi \circ \sigma_{L/K}^{\Omega_{K/\mathbf{Z}}^{\cdot}}.$$

In particular this map is independent of the chosen basis of Ω_K^1, and it induces the trace

$$\sigma_{L/K}^{\Omega_K} : \Omega_L \to \Omega_K$$

of [KD], §16. The remaining parts of (5.8) now follow immediately from (3.4) (T3).

5.9. Remark: Under the assumptions of (5.8) let R be local with maximal ideal $\mathfrak{m}$, let $k := R/\mathfrak{m}$ and let $\ell := S/\mathfrak{m}S$. Then $(\ell/k, \Omega_k)$ is in class II, and therefore there exists a trace $\sigma_{\ell/k}^{\Omega_k} : \Omega_\ell \to \Omega_k$. If $\sigma_{S/R}^{\Omega} : \Omega_S \to \Omega$ is the map constructed in (5.8), then the following diagram commutes:

$$\begin{array}{ccc}
\Omega_S & \longrightarrow & \Omega_\ell \\
\downarrow{\sigma_{S/R}^{\Omega}} & & \downarrow{\sigma_{\ell/k}^{\Omega_k}} \\
\Omega & \longrightarrow & \Omega_k.
\end{array}$$

In particular it holds:

If in addition S is local with maximal ideal $\mathfrak{M}$, then $\sigma^{\Omega}_{S/R}(\mathfrak{M}, d\mathfrak{M}) \subseteq (\mathfrak{m}, d\mathfrak{m})$

Proof: From the canonical exact sequence ([**KD**], (4.17))

$$\mathfrak{m}/\mathfrak{m}^2 \xrightarrow{\alpha} \Omega^1/\mathfrak{m}\Omega^1 \xrightarrow{\bar{\beta}} \Omega^1_k \to 0$$

with $\alpha\left(x + \mathfrak{m}^2\right) = dx + \mathfrak{m}\Omega^1$ it follows:

There exists a basis $\omega_1, \ldots, \omega_n$ of Ω^1 having the following properties:

$$\eta_1 := \bar{\beta}\left(\omega_1 + \mathfrak{m}\Omega^1\right), \ldots, \eta_t := \beta\left(\omega_t + \mathfrak{m}\Omega^1\right)$$

is a k–basis of Ω^1_k for some $t \leq n$, and $\omega_{t+1,\ldots,}\omega_n \in (\mathfrak{m}, d\mathfrak{m})$.

With respect to these bases form $d_i, \bar{d}_i (i = 1, \ldots, n)$ and $\delta_j, \bar{\delta}_j (j = 1, \ldots, t)$ as in the proof of (4.11) for class III. In the equations

$$\beta\left(\omega_i\right) = \sum_{j=1}^{t} r'_{i,j}\eta_j (i = 1, \ldots, n)$$

it holds in this case:

$$r'_{i,j} = \delta_{i,j}(i = 1, \ldots, n; j = 1, \ldots, t)$$

Therefore we get (using the notations of the proof of (4.11)):

$$\delta_{j_1} \cdot \ldots \cdot \delta_{j_p} \left(tr^{\mathbf{Z}}_{\ell/k}\left(\theta^{\cdot}_{\ell/\mathbf{Z}}(\omega')\right)\right)$$
$$= \sum_{i_1 < \cdots < i_p} r'^{j_1,\ldots,j_p}_{i_1,\ldots,i_p} \beta\left(d_{i_1} \cdot \ldots \cdot d_{i_p}\left(tr^{\mathbf{Z}}_{S/R}\left(\theta^{\cdot}_{S/\mathbf{Z}}(\omega)\right)\right)\right).$$

By (2.21)i) we also may calculate $\sigma^{\Omega_k}_{\ell/k}$ using the map

$$\delta^{\cdot}_{k/\mathbf{Z}} : H.(k/\mathbf{Z}) \to \Omega_k,$$

constructed using the basis $\eta_1, \ldots, \eta_t$ of Ω^1_k. This implies that

$$\sigma^{\Omega_k}_{\ell/k}(\omega') = \beta\left(\sigma^{\Omega}_{S/R}(\omega)\right)$$

as desired.

5.10. Remark:

i) Let $(S/R, \Omega)$ be in class III. Suppose that R is noetherian, that $\mathrm{Ass}(R) = \mathrm{Min}\,(R)$, and that S/R is generically a complete intersection. If d' is the differentiation of Ω_S, and if d is the differentiation of Ω, then it holds:

$$\sigma^{\Omega}_{S/R} \circ d' = d \circ \sigma^{\Omega}_{S/R}.$$

ii) Let $(S/R, \Omega)$ and $(T/S, \Omega_S)$ be in class I or III. Suppose that R is noetherian, that $\mathrm{Ass}(R) = \mathrm{Min}(R)$, and that T/S is generically a complete intersection. If Ω is torsion free, then it holds:

$$\sigma^{\Omega}_{T/R} = \sigma^{\Omega}_{S/R} \circ \sigma^{\Omega_S}_{T/S}.$$

Proof:

i) Let $K = Q(R)$ be the full ring of fractions of R, and let $L := K \otimes_R S = Q(S)$. Since $\Omega \to \Omega_K$ is injective, it suffices by (4.11) to show the claim for $(L/K, \Omega_K)$ instead of $(S/R, \Omega)$. By assumption L/K is a complete intersection. Because of (5.6)ii), claim i) therefore follows from the corresponding statement for traces in complete intersections ([**KD**], §16).

ii) As above we may assume by (4.11) that $\dim(R) = 0$. Since it suffices to show the claim locally in R we may in addition assume that R is local. Then S and T are finite direct products of local R–algebras, and by TR4 (direct products) and TR3 (base change) we may assume that S and T are local.

If $\mathbf{Q} \subseteq R$, then we also may assume $\Omega = \Omega_{R/\mathbf{Z}}$. Since T/S is a complete intersection, and since S is local and noetherian, it holds by (5.6)i) and (5.1):

$$\theta_{S/\mathbf{Z}} \circ \sigma^{\Omega_{S/\mathbf{Z}}}_{T/S} = tr^{\mathbf{Z}}_{T/S} \circ \theta_{T/\mathbf{Z}}.$$

The claim now is an easy consequence of (3.4) (T5).

If $(S/R, \Omega)$ and $(T/S, \Omega_S)$ are in class III, then let

$$\delta_{R/\mathbf{Z}} : H_{\cdot}(R/\mathbf{Z}) \to \Omega$$

and

$$\delta_{S/\mathbf{Z}} : H.(S/\mathbf{Z}) \to \Omega_S$$

be the corresponding maps of (2.21). For $\bar{\omega} \in \Omega_T$ let $\omega \in \Omega'_{T/\mathbf{Z}}$ be a representative of $\bar{\omega}$. Then it holds by definition

$$\sigma^{\Omega_S}_{T/S}(\bar{\omega}) = \delta_{S/\mathbf{Z}} \left(tr^{\mathbf{Z}}_{T/S} \left(\theta_{T/\mathbf{Z}}(\omega) \right) \right).$$

S is local and noetherian, and T/S is a complete intersection. By (5.6)ii) and (5.1) there exists an $\eta \in \Omega'_{S/\mathbf{Z}}$ whose image in Ω_S is $\sigma^{\Omega_S}_{T/S}(\omega)$, and such that $\theta_{S/\mathbf{Z}}(\eta) = tr^{\mathbf{Z}}_{T/S} \left(\theta_{T/\mathbf{Z}}(\omega) \right)$. Therefore it holds

$$\sigma^{\Omega}_{S/R} \left(\sigma^{\Omega_S}_{T/S}(\bar{\omega}) \right) = \delta_{R/\mathbf{Z}} \left(tr^{\mathbf{Z}}_{S/R} \left(\theta_{S/\mathbf{Z}}(\eta) \right) \right) = \delta_{R/\mathbf{Z}} \left(tr^{\mathbf{Z}}_{S/R} \left(tr^{\mathbf{Z}}_{T/S} \left(\theta_{T/\mathbf{Z}}(\omega) \right) \right) \right).$$

The claim now follows from (3.4) (T5).

The statements of (5.1) and (5.4) are useful tools for explicit calculations of residue symbols. It is not known whether the traces constructed in §4 have the trace property of Lipman [L₁] (4.6.2). In the remainder of this section we will use a result of E. Kunz and A. Kliegl to give an explicit example of a finite free algebra S/R for which there exists no R–homomorphism

$$\sigma_{S/R} : \Omega'_{S/\mathbf{Z}} \to \Omega'_{R/\mathbf{Z}}$$

making the following diagram commute:

$$
\begin{array}{ccc}
\Omega'_{S/\mathbf{Z}} & \xrightarrow{\ \theta_{S/\mathbf{Z}}\ } & H.(S/\mathbf{Z}) \\
{\scriptstyle \sigma_{S/R}} \downarrow & & \downarrow {\scriptstyle tr_{S/R}} \\
\Omega'_{R/\mathbf{Z}} & \xrightarrow{\ \theta_{R/\mathbf{Z}}\ } & H.(R/\mathbf{Z}).
\end{array}
$$

In the polynomial algebra $\mathbf{Z}[X_1, \ldots, X_9]$ consider the ideal $\mathfrak{J}$ generated by the 2×2 minors of the matrix

$$\begin{pmatrix} X_1 & X_2 & X_3 \\ X_4 & X_5 & X_6 \\ X_7 & X_8 & X_9 \end{pmatrix}.$$

Then the ring

$$R = \mathbf{Z}[X_1, \ldots, X_9]/\mathfrak{I} = \mathbf{Z}[x_1, \ldots, x_9]$$

is a domain of dimension 6 ([**EH**]), and $K = Q(R)$ is a field of characteristic 0. On the free R–module

$$S := R \oplus RY_1 \oplus RY_2 \oplus RY_3$$

define a commutative multiplication with the multiplication table

$$
\begin{aligned}
Y_1^2 &= -\,x_4 x_8 + (x_5 + x_7)Y_1 + x_4 Y_2 \\
Y_1 Y_2 &= x_1 Y_3 \\
Y_1 Y_3 &= x_3 x_4 + x_7 Y_3 \\
Y_2^2 &= x_1 x_3 + x_8 Y_2 - x_2 Y_3 \\
Y_2 Y_3 &= -\,x_2 x_6 + x_3 Y_1 \\
Y_3^2 &= -\,x_6 x_8 + x_9 Y_1 + x_2 Y_2
\end{aligned}
$$

Then S is an associative R–algebra, and $L := K \otimes_R S$ is a complete intersection over K. In particular a trace of differential forms

$$\sigma_{L/K} : \Omega_{L/\mathbf{Z}}^{\cdot} \to \Omega_{K/\mathbf{Z}}^{\cdot}$$

is defined ([**KD**], §16). However it holds

THEOREM ([**KD**], §16, EX 2). *There cannot exist a map*

$$\sigma_{S/R} : \Omega_{S/\mathbf{Z}}^{\cdot} \to \Omega_{R/\mathbf{Z}}^{\cdot}$$

such that the following diagram commutes

$$
\begin{array}{ccc}
\Omega_{S/\mathbf{Z}}^{\cdot} & \xrightarrow{\;\;\mathrm{can}\;\;} & \Omega_{L/\mathbf{Z}}^{\cdot} \\
\downarrow{\scriptstyle \sigma_{S/R}} & & \downarrow{\scriptstyle \sigma_{L/K}} \\
\Omega_{R/\mathbf{Z}}^{\cdot} & \xrightarrow{\;\;\mathrm{can}\;\;} & \Omega_{K/\mathbf{Z}}^{\cdot}
\end{array}
$$

Assume now that in this example there exists an R–homomorphism $\sigma_{S/R} : \Omega'_{S/\mathbf{Z}} \to \Omega'_{R/\mathbf{Z}}$ making the following diagram commute:

$$
\begin{array}{ccc}
\Omega'_{S/\mathbf{Z}} & \xrightarrow{\;\theta'_{S/\mathbf{Z}}\;} & H.(S/\mathbf{Z}) \\
{\scriptstyle \sigma_{S/R}}\downarrow & {\scriptstyle tr_{S/R}}\downarrow & \\
\Omega'_{R/\mathbf{Z}} & \xrightarrow[\;\theta'_{R/\mathbf{Z}}\;]{} & H.(R/\mathbf{Z})
\end{array}
\qquad (*)
$$

Let $N := R\backslash\{0\}$. Since $R/\mathbf{Z}$ is flat we have isomorphisms $H.(R/\mathbf{Z})_N = H.(K/\mathbf{Z})$ and $H.(S/\mathbf{Z})_N = H.(L/\mathbf{Z})$, and by (3.4) (T3) we get $(tr_{S/R})_N = tr_{L/K}$. The functoriality of θ' therefore implies that $(*)$ induces a commutative diagram

$$
\begin{array}{ccc}
\Omega'_{L/\mathbf{Z}} & \xrightarrow{\;\theta'_{L/\mathbf{Z}}\;} & H.(L/\mathbf{Z}) \\
{\scriptstyle (\sigma_{S/R})_N}\downarrow & {\scriptstyle tr_{L/K}}\downarrow & \\
\Omega'_{K/\mathbf{Z}} & \xrightarrow[\;\theta'_{L/\mathbf{Z}}\;]{} & H.(K/\mathbf{Z})
\end{array}
\qquad (**)
$$

On the other hand (5.1) implies that also the following diagram commutes

$$
\begin{array}{ccc}
\Omega'_{L/\mathbf{Z}} & \xrightarrow{\;\theta'_{L/\mathbf{Z}}\;} & H.(L/\mathbf{Z}) \\
{\scriptstyle \sigma_{L/K}}\downarrow & {\scriptstyle tr_{L/K}}\downarrow & \\
\Omega'_{K/\mathbf{Z}} & \xrightarrow[\;\theta'_{K/\mathbf{Z}}\;]{} & H.(K/\mathbf{Z})
\end{array}
\qquad (***)
$$

Since $\theta'_{K/\mathbf{Z}}$ is an isomorphism by (2.23) $(**)$ and $(***)$ imply that $\sigma_{L/K} = (\sigma_{S/R})_N$ which in turn implies the commutativity of the following diagram

$$
\begin{array}{ccc}
\Omega'_{S/\mathbf{Z}} & \xrightarrow[can]{} & \Omega'_{L/\mathbf{Z}} \\
{\scriptstyle \sigma_{S/R}}\downarrow & {\scriptstyle \sigma_{L/K}}\downarrow & \\
\Omega'_{R/\mathbf{Z}} & \xrightarrow[can]{} & \Omega'_{K/\mathbf{Z}}
\end{array}
$$

in contradiction to the theorem mentioned above.

§6. The topological residue homomorphism.

In this section we will use the theory of topological Hochschild homology and Hochschild cohomology to generalize Lipman's definition of residues via Hochschild homology to certain classes of topological algebras. All rings in this section are assumed to be noetherian.

Let $(R, \tau)/k$ be a topological algebra and suppose that τ is the $\mathfrak{J}$-adic topology on R for some ideal $\mathfrak{J} \subseteq R$. Let $P := R/\mathfrak{J}$ and assume that P/k is finite and projective as a k-module. Note that under these assumptions $R \hat{\otimes}_k R$ is a noetherian ring by [**AM**], (10.25). Furthermore $\operatorname{Hom}_k(P, P)$ has an obvious $R \hat{\otimes}_k R$-module structure, and it is complete in its $\tau \hat{\otimes} \tau$-adic topology since it is a finite $R \hat{\otimes}_k R$-module. Therefore the topological Hochschild cohomology $H^{\cdot}(R/k, \tau, \operatorname{Hom}_k(P, P))$ with coefficients in $\operatorname{Hom}_k(P, P)$ is well defined, and it comes equipped with an associative graded R-algebra structure.

Define the residue homomorphism

$$\operatorname{Res}^q_{R/k,\tau,P} : H_q(R/k, \tau) \otimes_R H^q(R/k, \tau, \operatorname{Hom}_k(P, P)) \to k$$

in the following way:

We may assume that R is complete in its τ-adic topology. Let $\xi \in H_q(R/k, \tau)$ be represented by $x \in \gamma_q(R/k, \tau)$ and let $\eta \in H^q(R/k, \tau, \operatorname{Hom}_k(P, P))$ be represented by $f \in \operatorname{Mult}^q_{k,\mathrm{cont}}(R, \operatorname{Hom}_k(P, P)) \cong \operatorname{Hom}_{k,\mathrm{cont}}(T^q_k(R, \tau), \operatorname{Hom}_k(P, P))$. Then f induces

$$\tilde{f} \in \operatorname{Hom}_{\widehat{R^e}}(\widehat{R^e} \hat{\otimes}_k T^q_k(R, \tau), \operatorname{Hom}_k(P, P)) = \operatorname{Hom}_{\widehat{R^e}}(\beta_q(R/k, \tau), \operatorname{Hom}_k(P, P))$$

and therefore we get an induced map

$$\overline{f} := 1 \otimes \tilde{f} : R \widehat{\otimes_{\widehat{R^e}}} \beta_q(R/k, \tau) = \gamma_q(R/k, \tau) \to R \otimes_{\widehat{R^e}} \operatorname{Hom}_k(P, P),$$

using (1.12)ii).

By [**EGA O$_I$**] (7.7.8) we get

$$R \widehat{\otimes_{\widehat{R^e}}} \operatorname{Hom}_k(P, P) \cong R \otimes_{R^e} \operatorname{Hom}_k(P, P)$$
$$\cong \operatorname{Hom}_k(P, P)/\langle\{r\varphi - \varphi r : r \in R, \ \varphi \in \operatorname{Hom}_k(P, P)\}\rangle$$
$$\cong H_0(R, \operatorname{Hom}_k(P, P))$$

and therefore we can define:

$$\text{Res}^q_{R/k,\tau,P}(\xi \otimes \eta) := Tr_{P/k}(\overline{f}(x)).$$

Since $\overline{\delta g}(y) = \pm\overline{g}(dy)$, where δ is the coboundary map of $\underset{n \in \mathbb{N}}{\oplus} \text{Mult}^n_{k,\text{cont}}(R, \text{Hom}_k(P,P))$, and where d is the boundary map of $\gamma.(R/k,\tau)$, this definition depends on ξ and η only.

6.1. Examples:

i. ($q = 0$) As above assume that R is complete. Then we have $H_0(R/k,\tau) = R$ and $H^0(R/k,\tau,\text{Hom}_k(P,P)) = \{f \in \text{Hom}_k(P,P) : rf = fr \text{ for all } r \in R\} = \text{Hom}_R(P,P) \subsetneq \text{Hom}_k(P,P)$. Thus for $f \in \text{Hom}_R(P,P), r \in R$ we have as in the discrete case

$$\text{Res}^0_{R/k,\tau,P}(r \otimes f) = Tr_{P/k}(rf) = Tr_{P/k}(fr).$$

ii. Let $\eta \in H^q(R/k,\tau,\text{Hom}_k(P,P))$ be represented by $f \in \text{Mult}^q_{k,\text{cont}}(R, \text{Hom}_k(P,P))$ and let $\xi \in H_q(R/k,\tau)$ be represented by $x \in \gamma_q(R/k,\tau)$. Then we can write x as a convergent series $\Sigma r_0^{(i)} \hat{\otimes} r_1^{(i)} \hat{\otimes} \ldots \hat{\otimes} r_q^{(i)}$ in the canonical topology of $\gamma_q(R/k,\tau)$. Then we have

$$\text{Res}^q_{R/k,\tau,P}(\xi \otimes \eta) = Tr_{P/k}(\sum r_0^{(i)} f(r_1^{(i)}, \ldots, r_q^{(i)})).$$

iii. ($q = 1$) Again we may assume that R is complete. Let σ be a k–linear section of the natural map $\pi : R/\mathfrak{I}^2 \to R/\mathfrak{I}$ and set

$$(\mathfrak{I}/\mathfrak{I}^2)^* = \text{Hom}_P(\mathfrak{I}/\mathfrak{I}^2, P).$$

Since τ is the $\mathfrak{I}$–adic topology we have a canonical isomorphism

$$H^1(R/k,\tau,\text{Hom}_k(P,P)) \cong H^1(R/k,\text{Hom}_k(P,P)),$$

and therefore we get as in the discrete case an isomorphism

$$H^1(R/k,\tau,\text{Hom}_k(P,P)) \tilde{\to} (\mathfrak{I}/\mathfrak{I}^2)^*$$

(see c.f. [L_1], §1). Furthermore

$$\theta^1_{(R/k,\tau)} : \widetilde{\Omega}^1_{R/k} \to H.(R/k,\tau)$$

is an isomorphism by (2.7) and (2.9), and calculations similar to those in [**L₁**], §1 show that

$$\mathrm{Res}^1_{R/k,\tau,P} : \tilde{\Omega}^1_{R/k} \otimes_R (\mathfrak{I}/\mathfrak{I}^2)^* \to k$$

is given by

$$\mathrm{Res}^1_{R/k,\tau,P}(\alpha \otimes dr) = Tr_{P/k}(\alpha \circ (r\sigma - \sigma r)).$$

The topological residue homomorphism is closely related to Lipman's construction. Recall that there are canonical maps

$$i : H^\cdot(R/k, \tau, \mathrm{Hom}_k(P,P)) \to H^\cdot(R/k, \mathrm{Hom}_k(P,P))$$

$$p : H_\cdot(R/k) \to H_\cdot(R/k, \tau).$$

With this notation we get

6.2. PROPOSITION. *The following diagram commutes*

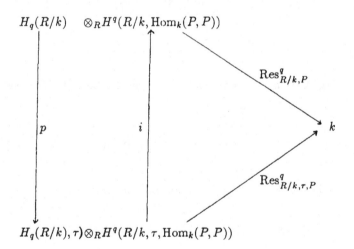

i.e. for $\xi \in H_q(R/k)$ and $\eta \in H^q(R/k, \tau, \mathrm{Hom}_k(P,P))$ it holds

$$\mathrm{Res}^q_{R/k,P}(\xi \otimes i(\eta)) = \mathrm{Res}^q_{R/k,\tau,P}(p(\xi) \otimes \eta).$$

Proof: Let ξ be represented by $\Sigma r_0^{(i)} \otimes r_1^{(i)} \otimes \cdots \otimes r_q^{(i)} \in \gamma_q(R/k)$, and let η be represented by $f \in \mathrm{Mult}^q_{k,\mathrm{cont}}(R, \mathrm{Hom}_k(P,P))$. Then $p(\xi)$ is represented by $\Sigma r_0^{(i)} \hat{\otimes} r_1^{(i)} \hat{\otimes} \ldots \hat{\otimes} r_q^{(i)} \in$

$\gamma_q(R/k, \tau)$, and $i(\eta)$ is represented by f, considered as an element of $\mathrm{Mult}_k^q(R, \mathrm{Hom}_k(P, P))$.
By (6.1)ii. and $[\mathbf{L_1}]$ (1.5.2) we get:

$$\mathrm{Res}_{R/k,P}^q(\xi \otimes i(\eta)) = Tr_{P/k}\left(\sum r_0^{(i)} f(r_1^{(i)}, \ldots, r_q^{(i)})\right)$$
$$= \mathrm{Res}_{R/k,\tau,P}^q(p(\xi) \otimes \eta),$$

proving the proposition.

As was mentioned in (6.1) iii. one has a canonical isomorphism

$$(\mathfrak{I}/\mathfrak{I}^2)^* = \mathrm{Hom}_P(\mathfrak{I}/\mathfrak{I}^2, P) \to H^1(R/k, \tau, \mathrm{Hom}_k(P, P))$$

which by the universal property of the tensor algebra induces a well defined homomorphism
of graded R–algebras

$$\bigoplus_{n \geq 0} T_R^n((\mathfrak{I}/\mathfrak{I}^2)^*) \to H^{\cdot}(R/k, \tau, \mathrm{Hom}_k(P, P)).$$

Similarly Lipman in $[\mathbf{L_1}]$, (1.8.3) gets a homomorphism

$$\bigoplus_{n \geq 0} T_R^n((\mathfrak{I}/\mathfrak{I}^2)^*) \to H^{\cdot}(R/k, \mathrm{Hom}_k(P, P)),$$

and by the very definitions the following diagram commutes:

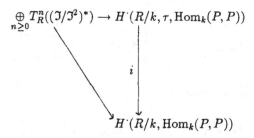

Denoting for $\alpha_1, \ldots, \alpha_q \in (\mathfrak{I}/\mathfrak{I}^2)^*$ by $[\alpha_1 \ldots \alpha_q]$ the image of $\alpha_1 \otimes \cdots \otimes \alpha_q$ by the above map,
and by $\mathrm{Res}_{R/k,\tau}^q \begin{bmatrix} \\ \alpha_1, \ldots, \alpha_q \end{bmatrix}$ the map $\mathrm{Res}_{R/k,\tau,P}^q(\_ \otimes [\alpha_1 \ldots \alpha_q])$ we get as an immediate
consequence of (6.2):

6.3. COROLLARY. *For $\alpha_1, \ldots, \alpha_q \in (\mathfrak{I}/\mathfrak{I}^2)^*$ the following diagram commutes*

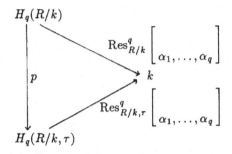

Lipman in [L₁] (1.10) uses the canonical homomorphism

$$\theta^{\cdot}_{R/k} : \Omega^{\cdot}_{R/k} \to H.(R/k)$$

to define for $\omega \in \Omega^q_{R/k}$:

$$\mathrm{Res}^q_{R/k} \begin{bmatrix} \omega \\ \alpha_1, \ldots, \alpha_q \end{bmatrix} := \mathrm{Res}^q_{R/k} \begin{bmatrix} \theta^{\cdot}_{R/k}(\omega) \\ \alpha_1, \ldots, \alpha_q \end{bmatrix}.$$

If under the present assumptions R is complete in its τ–adic topology, then the universally finite differential algebra $\tilde{\Omega}^{\cdot}_{R/k}$ of R/k exists, and it is canonically isomorphic to the universally complete differential algebra $\Omega^{\cdot}_{(R/k,\tau)}$ by (2.6)ii). Therefore we can use the canonical homomorphism

$$\theta^{\cdot}_{(R/k,\tau)} : \tilde{\Omega}^{\cdot}_{R/k} \to H.(R/k,\tau)$$

to define for differential forms $\tilde{\omega} \in \tilde{\Omega}^q_{R/k}$:

$$\widetilde{\mathrm{Res}}^q_{R/k} \begin{bmatrix} \tilde{\omega} \\ \alpha_1, \ldots, \alpha_q \end{bmatrix} = \mathrm{Res}^q_{R/k,\tau} \begin{bmatrix} \theta^{\cdot}_{(R/k,\tau)}(\tilde{\omega}) \\ \alpha_1, \ldots, \alpha_q \end{bmatrix}.$$

The functoriality of $\theta^{\cdot}$ in connection with (6.3) implies:

6.4. COROLLARY. *For $\omega \in \Omega^q_{R/k}$ denote by $\tilde{\omega}$ its image in $\tilde{\Omega}^q_{R/k}$ by the canonical map $\Omega^{\cdot}_{R/k} \to \tilde{\Omega}^{\cdot}_{R/k}$. Then*

$$\widetilde{\mathrm{Res}}^q_{R/k} \begin{bmatrix} \tilde{\omega} \\ \alpha_1, \ldots, \alpha_q \end{bmatrix} = \mathrm{Res}^q_{R/k} \begin{bmatrix} \omega \\ \alpha_1, \ldots, \alpha_q \end{bmatrix}.$$

6.5. Remark: Suppose that $\mathfrak{I}$ is generated by an R–regular sequence $f_1, \ldots, f_q$. Then the P–module $\mathfrak{I}/\mathfrak{I}^2$ is free with basis $\overline{f}_1, \ldots, \overline{f}_q$, where $\overline{f}_i$ denotes the natural image of f_i in $\mathfrak{I}/\mathfrak{I}^2$. Let $\alpha_1, \ldots, \alpha_q$ be the basis of $(\mathfrak{I}/\mathfrak{I}^2)^*$ dual to the basis $\overline{f}_1, \ldots, \overline{f}_q$ of $\mathfrak{I}/\mathfrak{I}^2$. Then we set for $\xi \in H_q(R/k, \tau)$

$$\mathrm{Res}^q_{R/k,\tau} \begin{bmatrix} \xi \\ f_1, \ldots, f_q \end{bmatrix} = \mathrm{Res}^q_{R/k,\tau} \begin{bmatrix} \xi \\ \alpha_1, \ldots, \alpha_q \end{bmatrix},$$

and similarly we set for $\widetilde{\omega} \in \widetilde{\Omega}^q_{R/k}$

$$\widetilde{\mathrm{Res}}^q_{R/k} \begin{bmatrix} \widetilde{\omega} \\ f_1, \ldots, f_q \end{bmatrix} = \widetilde{\mathrm{Res}}^q_{R/k} \begin{bmatrix} \widetilde{\omega} \\ \alpha_1, \ldots, \alpha_q \end{bmatrix}.$$

6.6. Remark: The formula in (6.4) shows that Lipman's results for the residue symbol as derived in [$\mathbf{L_1}$]hold for the topological residue symbol as well. In particular the formulas in [$\mathbf{L_1}$], §3 can be used for explicit calculations.

In the remainder of this section we will show how to use traces of differential forms to reduce the calculations of residues to power series algebras, if the ideal $\mathfrak{I}$ is generated by a regular sequence.

Let k be a local ring, let $R = k[[X_1, \ldots, X_n]]$ be a power series algebra over k, and let S/R be a finite flat R-algebra. By abuse of notation we denote by τ both the $(X_1, \ldots, X_n)$–adic topology on R and on S. Recall that by (2.6) and (2.7) the universally finite differential algebras $\widetilde{\Omega}_{R/k}$ of R/k and $\widetilde{\Omega}_{S/k}$ of S/k exist and are canonically isomorphic to the universally complete differential algebras. By [**KD**], (11.9) we have

$$\widetilde{\Omega}^{\cdot}_{S/k} = (\widetilde{\Omega}^{\cdot}_{R/k})_S$$

where as usual $(\widetilde{\Omega}^{\cdot}_{R/k})_S$ denotes the universal S–extension of $\widetilde{\Omega}^{\cdot}_{R/k}$, and by (2.14) the canonical map

$$\theta_{(R/k,\tau)} : \widetilde{\Omega}^{\cdot}_{R/k} \cong \Omega^{\cdot}_{(R/k,\tau)} \to H.(R/k, \tau)$$

is an isomorphism of DG–algebras.

Therefore we get a well defined trace map

$$\widetilde{\sigma}_{S/R} : \widetilde{\Omega}^{\cdot}_{S/k} \to \widetilde{\Omega}^{\cdot}_{R/k}$$

given by the formula

$$\tilde{\sigma}_{S/R} = (\theta'_{(R/k,\tau)})^{-1} \circ tr_{(S/R,\tau)} \circ \theta'_{(S/k,\tau)}$$

where $tr_{(S/R,\tau)}$ is the trace constructed in §3.

6.7. Remark:

i. The following diagram commutes

$$
\begin{array}{ccc}
\tilde{\Omega}^{\cdot}_{S/k} & \xrightarrow{\;\theta'_{(S/k,\tau)}\;} & H.(S/k,\tau) \\
\downarrow{\scriptstyle \tilde{\sigma}_{S/R}} & {\scriptstyle tr_{(S/R,\tau)}}\downarrow & \\
\tilde{\Omega}^{\cdot}_{R/k} & \xrightarrow{\;\theta'_{(R/k,\tau)}\;} & H.(R/k,\tau).
\end{array}
$$

ii. Under the above assumptions $\tilde{\Omega}^1_{R/k}$ is free of rank n. Therefore we also get a trace map by the construction of (4.4). These two maps coincide, because the inverse of $\theta'_{(R/k,\tau)}$ can be described in terms of derivations similar to the description of $\delta'_{R/k}$ in (2.20). The commutativity of the diagram in i. shows that in this special case the construction in (4.4) is independent of the choice of a basis of $\tilde{\Omega}^1_{R/k}$.

iii. If $\mathbf{Q}$ is contained in R, then $\tilde{\sigma}_{S/R}$ coincides with the trace morphism constructed in (4.2).

iv. If S/R is a complete intersection, then $\tilde{\sigma}_{S/R}$ coincides with the trace map of [**KD**], §16. In view of i. this follows easily from (5.5).

v. The map $\tilde{\sigma}_{S/R}$ satisfies the trace axioms TR1, TR2, TR4, TR6 and TR7.

Let $\mathfrak{I} \subseteq R$ be an ideal such that $\mathfrak{I}$ defines the topology τ on R and S. Denote by $\mathfrak{I}' := \mathfrak{I}S$, and suppose that $P := R/\mathfrak{I}$ is finite and flat as a k–module. Let $P' := S/\mathfrak{I}'$, and for $\alpha \in \mathrm{Hom}_P(\mathfrak{I}/\mathfrak{I}^2, P)$ denote by $\alpha' \in \mathrm{Hom}_{P'}(\mathfrak{I}'/(\mathfrak{I}')^2, P')$ its cotrace in the sense of [**L₁**], (4.4).

6.8. PROPOSITION. For $\alpha_1, \ldots, \alpha_q \in \mathrm{Hom}_P(\mathfrak{I}/\mathfrak{I}^2, P)$ and $\tilde{\omega} \in \Omega^q_{S/k}$ it holds

$$
\widetilde{\mathrm{Res}}^{\,q}_{S/k}
\begin{bmatrix} \tilde{\omega} \\ \alpha'_1, \ldots, \alpha'_q \end{bmatrix}
= \widetilde{\mathrm{Res}}^{\,q}_{R/k}
\begin{bmatrix} \tilde{\sigma}_{S/R}(\tilde{\omega}) \\ \alpha_1, \ldots, \alpha_q \end{bmatrix}.
$$

Proof: By the assumptions we can use the topology τ to calculate the residues in question. One gets a diagram

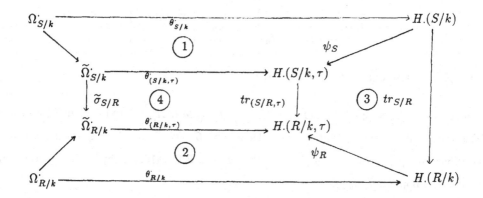

In which the subdiagrams $\textcircled{1}$, $\textcircled{2}$, $\textcircled{3}$ and $\textcircled{4}$ commute by (2.10), (3.4) (T3) and (3.1). So if $\omega \in \Omega^q_{S/k}$ is a preimage of $\tilde{\omega} \in \tilde{\Omega}^q_{S/k}$ it holds:

$$\widetilde{\mathrm{Res}}^q_{S/k}\begin{bmatrix} \tilde{\omega} \\ \alpha'_1, \ldots, \alpha'_q \end{bmatrix}$$

$$= \mathrm{Res}^q_{(S/k,\tau)}\begin{bmatrix} \theta'_{(S/k,\tau)}(\tilde{\omega}) \\ \alpha'_1, \ldots, \alpha'_q \end{bmatrix}$$

$$= \mathrm{Res}^q_{S/k}\begin{bmatrix} \theta'_{S/k}(\omega) \\ \alpha'_1, \ldots, \alpha'_q \end{bmatrix} \qquad \text{(by (6.3))}$$

$$= \mathrm{Res}^q_{R/k}\begin{bmatrix} tr_{S/R}(\theta'_{S/k}(\omega)) \\ \alpha_1, \ldots, \alpha_q \end{bmatrix} \qquad \text{([L}_1\text{], (4.7.2))}$$

$$= \mathrm{Res}^q_{(R/k,\tau)}\begin{bmatrix} \psi_R(tr_{S/R}(\theta'_{S/k}(\omega))) \\ \alpha_1, \ldots, \alpha_q \end{bmatrix} \qquad \text{(by (6.3))}$$

$$= \mathrm{Res}^q_{(R/k,\tau)}\begin{bmatrix} tr_{(S/R,\tau)}(\psi_S(\theta'_{S/k}(\omega))) \\ \alpha_1, \ldots, \alpha_q \end{bmatrix} \qquad (\textcircled{3} \text{ commutes})$$

$$= \mathrm{Res}^q_{(R/k,\tau)}\begin{bmatrix} \theta'_{(R/k,\tau)}(\tilde{\sigma}_{S/R}(\tilde{\omega})) \\ \alpha_1, \ldots, \alpha_q \end{bmatrix} \qquad (\textcircled{1} \text{ and } \textcircled{4} \text{ commute})$$

$$= \widetilde{\mathrm{Res}}^q_{R/k}\begin{bmatrix} \tilde{\sigma}_{S/R}(\tilde{\omega}) \\ \alpha_1, \ldots, \alpha_q \end{bmatrix}$$

which is the claim of the proposition.

6.9. COROLLARY. *If* $\{f_1, \ldots, f_n\} \subseteq (X_1, \ldots, X_n)R$ *is a regular sequence of* R *such that* $R/(f_1, \ldots, f_n)$ *is finite and flat as a* k–*module, then it holds for* $\widetilde{\omega} \in \widetilde{\Omega}^n_{S/k}$:

$$\widetilde{\operatorname{Res}}^n_{S/k} \begin{bmatrix} \widetilde{\omega} \\ f_1, \ldots, f_n \end{bmatrix} = \widetilde{\operatorname{Res}}^n_{R/k} \begin{bmatrix} \widetilde{\sigma}_{S/R}(\widetilde{\omega}) \\ f_1, \ldots, f_n \end{bmatrix}$$

where by f_i *we also denote the image of* f_i *in* S.

6.10. *Remark:* Suppose in addition that $\operatorname{Ass}(R) = \operatorname{Min}(R)$, $\operatorname{Ass}(S) = \operatorname{Min}(S)$, and that S/R is generically a complete intersection. Denote by $L = Q(S)$ and $K = Q(R)$ the full rings of fractions, by $\sigma_{L/K} : (\widetilde{\Omega}_{S/k})_L \to (\widetilde{\Omega}_{R/k})_K$ the trace of [**KD**], §16, and by $i : \widetilde{\Omega}_{S/k} \to (\widetilde{\Omega}_{S/k})_L$ the canonical homomorphism.

Identifying $\widetilde{\Omega}_{R/k}$ with its image in $(\widetilde{\Omega}_{R/k})_K$ we get in view of (5.8) for $f_1, \ldots, f_n$ as in (6.9) and for $\widetilde{\omega} \in \widetilde{\Omega}_{S/k}$:

$$\widetilde{\operatorname{Res}}^n_{S/k} \begin{bmatrix} \widetilde{\omega} \\ f_1, \ldots, f_n \end{bmatrix} = \widetilde{\operatorname{Res}}^n_{R/k} \begin{bmatrix} \sigma_{L/K}(i(\widetilde{\omega})) \\ f_1, \ldots, f_n \end{bmatrix}.$$

In particular it follows that the residue symbol is independent of the torsion of $\widetilde{\Omega}_{S/k}$.

6.11. *Example:* Let S/k be an algebra, and let $\mathfrak{J} \subseteq S$ be an ideal such that $P := S/\mathfrak{J}$ is finite and flat as a k–module and such that S is complete in its $\mathfrak{J}$–adic topology. Furthermore suppose that $\mathfrak{J}$ is generated by an S–regular sequence $f_1, \ldots, f_n$. Then S is a finite and flat module over $R = k[[X_1, \ldots, X_n]]$ via $X_i \mapsto f_i$ by ([**L**$_1$], (3.3.2)), and the formula (6.9) can be used to calculate residues and to reduce the problem to R.

Suppose now that in addition k and S are reduced and that $\mathbf{Q} \subseteq k$. Using the notation of (6.10) we have that L/K is étale, and therefore we can write $i(\widetilde{\omega}) = \ell df_1 \ldots df_n$ for a suitable $\ell \in L$. Then (6.10) implies

$$\widetilde{\operatorname{Res}}^n_{S/k} \begin{bmatrix} \widetilde{\omega} \\ f_1, \ldots, f_n \end{bmatrix} = \sigma_{L/K}(\ell)(0)$$

where for $r \in R$ we denote by $r(0)$ the image of r in $R/(X_1, \ldots, X_n) \cong k$.

6.12. *Example:* Suppose $S = k[[X_1, \ldots, X_n]]$ and $\mathfrak{J} = (X_1^{m_1}, \ldots, X_n^{m_n})$ for some $m_i > 0$, and set $R = k[[X_1^{m_1}, \ldots, X_n^{m_n}]]$. Then S/R is a complete intersection, and therefore the trace of [**KD**], §16 can be used to calculate residues. If we write $\widetilde{\omega} = f dX_1 \ldots dX_n$ for some $f = \Sigma a_{\mu_1,\ldots,\mu_n} X_1^{\mu_1} \ldots X_n^{\mu_n}$ in R, then it follows from [**KD**], (16.6):

$$\widetilde{\mathrm{Res}}_{S/k}^n \begin{bmatrix} \widetilde{\omega} \\ X_1^{m_1}, \ldots, X_n^{m_n} \end{bmatrix} = a_{m_1-1,\ldots,m_n-1}$$

6.13. Remark: The result of (6.9) can also be used to give a different proof of the compatibility of residues with exterior differentiation ([**L₁**], appendix B) in case k is noetherian. In fact in order to show

$$\mathrm{Res}_{S/k}^n \begin{bmatrix} d\omega \\ f_1^{m_1}, \ldots, f_n^{m_n} \end{bmatrix} = \sum_{k=1}^n m_k \, \mathrm{Res}_{S/k}^n \begin{bmatrix} df_k \cdot \omega \\ f_1^{m_1}, \ldots, f_k^{m_k+1}, \ldots, f_n^{m_n} \end{bmatrix}$$

for $\omega \in \Omega_{S/k}^{n-1}$ and positive integers $m_1, \ldots, m_n$, we may assume that S is complete in its $(f_1, \ldots, f_n)$-adic topology. Then S is noetherian as well by [**AM**], (10.25). By (6.4) it suffices to show this formula for the toplogical residue homomorphism. Denoting again by $R = k[[X_1, \ldots, X_n]]$, we can reduce to the case $S = R$ and $f_i = X_i$ by (6.9), since $\widetilde{\sigma}_{S/R}$ commutes with differentiation and is $\widetilde{\Omega}_{R/k}$-linear. The formula now follows easily from (6.12).

§7. Trace formulas for residues of differential forms.

Traces in Hochschild homology are an important tool to calculate residues in Hochschild homology via "trace formula II" ([L₁], (4.7.1)). Lipman also shows that suitable traces of differential forms give rise to a similar trace formula for residues of differential forms ([L₁], (4.7.3)):

Let R/k be an algebra, let S/R be a finite projective algebra, and suppose that there exists a trace of differential forms

$$\sigma_{S/R} : \Omega^{\cdot}_{S/k} \to \Omega^{\cdot}_{R/k}$$

which makes the following diagram commute

$$
\begin{array}{ccc}
\Omega^{\cdot}_{S/k} & \xrightarrow{\ \theta^{\cdot}_{S/k}\ } & H_{\cdot}(S/k) \\
\downarrow{\scriptstyle \sigma_{S/R}} & {\scriptstyle tr_{S/R}}\downarrow & \\
\Omega^{\cdot}_{R/k} & \xrightarrow{\ \theta^{\cdot}_{R/k}\ } & H_{\cdot}(R/k).
\end{array}
\qquad (*)
$$

Furthermore let $\mathfrak{I} \subseteq R$ be an ideal of R such that $P := R/\mathfrak{I}$ is a finite projective k–module, let $\mathfrak{I}' := \mathfrak{I}S$, and let $P' := S/\mathfrak{I}' = P \otimes_R S$. Then there exist injective maps ([L₁], (1.4))

$$\overline{\psi} : H^1(R/k, \operatorname{Hom}_k(P, P)) \to \operatorname{Hom}_P(\mathfrak{I}/\mathfrak{I}^2, P)$$

and

$$\overline{\psi'} : H^1(S/k, \operatorname{Hom}_k(P', P')) \to \operatorname{Hom}_{P'}(\mathfrak{I}'/\mathfrak{I}'^2, P')$$

and a canonical cotrace ([L₁], (4.3))

$$\gamma^1 : H^1(R/k, \operatorname{Hom}_k(P, P)) \to H^1(S/k, \operatorname{Hom}_k(P', P')).$$

for $\xi_1, \ldots, \xi_q \in H^1(R/k, \operatorname{Hom}_k(P, P))$ let

$$\alpha_i = \overline{\psi}(\xi_i) \in \operatorname{Hom}_P(\mathfrak{I}/\mathfrak{I}^2, P) \qquad 1 \leq i \leq q$$

and

$$\alpha'_i = \overline{\psi'}(\gamma^1(\xi_i)) \in \operatorname{Hom}_{P'}(\mathfrak{I}'/\mathfrak{I}'^2, P') \qquad 1 \leq i \leq q.$$

Under these assumptions it holds for $\omega \in \Omega^q_{S/k}$

$$\operatorname{Res}^q_{S/k} \begin{bmatrix} \omega \\ \alpha'_1, \ldots, \alpha'_q \end{bmatrix} = \operatorname{Res}^q_{R/k} \begin{bmatrix} \sigma_{S/R}(\omega) \\ \alpha_1, \ldots, \alpha_q \end{bmatrix}.$$

The results of §5 show that this formula holds for the trace $\sigma_{S/R}$ of [KD], §16 if S/R is locally a complete intersection and if either R is local or R/k is flat. It is not known whether the traces constructed in §4 give rise to a commutative diagram (*), however we will show that the above formula still is valid, if k is reduced, if $\mathfrak{I}$ is generated by an R–regular sequence $f_1, \ldots, f_n$, and if $\alpha_1, \ldots, \alpha_n$ is the basis of $\operatorname{Hom}_P(\mathfrak{I}/\mathfrak{I}^2, P)$ dual to the basis $f_1 + \mathfrak{I}^2, \ldots, f_n + \mathfrak{I}^2$ of $\mathfrak{I}/\mathfrak{I}^2$. This trace formula will be used to deduce residue axiom (R4) "transitivity" ([RD], p.199) which is the only residue formula in [RD] not proved in [L1].

Before starting with the proof of "trace formula II" we need the following lemma:

7.1. LEMMA. *Let R be a noetherian ring and let S/R be a noetherian algebra. Suppose that $f_1, \ldots, f_n \in S$ are elements for which $Q = S/(f_1, \ldots, f_n)$ is finite as an R-module. Furthermore let $\mathfrak{I} \subseteq R$ be an ideal and denote by $\hat{R}$ the $\mathfrak{I}$-adic completion of R, by $\hat{S}$ the $(\mathfrak{I}, f_1, \ldots, f_n)$-adic completion of S, and by $\hat{Q}$ the $\hat{S}$-module $\hat{S}/(f_1, \ldots, f_n)\hat{S}$.*

Then the canonical homomorphism

$$Q \otimes_R \hat{R} \to \hat{Q}$$

is an isomorphism.

Proof: By ([Mat], (23.I), Prop.) we have a canonical isomorphism $\hat{S}/(f_1, \ldots, f_n)\hat{S} = (S/(f_1, \ldots, f_n)S)\hat{\ }$ where the latter $\hat{\ }$ denotes the $(\mathfrak{I}, f_1, \ldots, f_n)S$-adic completion of the S-module $Q = S/(f_1, \ldots, f_n)S$. But clearly the $(\mathfrak{I}, f_1, \ldots, f_n)$ S-adic topology on Q is the same as the $\mathfrak{I}S$-adic topology on Q, which in turn is the same as the $\mathfrak{I}R$-adic topology on the finite R-module Q. From this and from R being noetherian the claim follows (c.f. [Mat], thm. 55).

Now let again k be a noetherian ring, and let R/k be a noetherian algebra. Suppose that $\{f_1, \ldots, f_d\} \subseteq R$ is a quasi-regular sequence such that $R/(f_1, \ldots, f_d)$ is finite and flat as a k-module, and let S/R be a finite flat algebra. Using the notation introduced in (6.5) we get:

7.2. THEOREM ("TRACE FORMULA II"). *Suppose that one of the following three conditions holds:*

i. S/R *is a finite locally complete intersection.*

ii. k *is a reduced* **Q**–*algebra.*

iii. k *is a reduced ring,* 2 *is not a zerodivisor of* R *and* $\Omega^1_{R/k}$ *is finite and projective as an* R–*module.*

Then we have a well defined trace map

$$\sigma_{S/R} : \Omega^{\cdot}_{S/k} \to \Omega^{\cdot}_{R/k}$$

and it holds for $\omega \in \Omega^d_{S/k}$

$$\operatorname{Res}^d_{S/k}\left[\begin{array}{c}\omega \\ f_1,\ldots,f_d\end{array}\right] = \operatorname{Res}^d_{R/k}\left[\begin{array}{c}\sigma_{S/R}(\omega) \\ f_1,\ldots,f_d\end{array}\right]$$

If the universally finite differential algebra $\widetilde{\Omega}_{R/k}$ of R/k exists (and is projective in case iii.), then $\widetilde{\Omega}_{S/k}$ exists as well, and $\widetilde{\Omega}_{S/k} = (\widetilde{\Omega}_{R/k})_S$ by [**KD**], (11.9). In this case the above formula also holds true for universally finite differential forms and their traces.

Proof: In any case it suffices to prove the formula locally in k since both residues and traces commute with base change. Therefore we may assume k is local with maximal ideal $\mathfrak{m}$.

Under assumption i. we also may replace k by its $\mathfrak{m}$–adic completion and R and S by their $(\mathfrak{m}, f_1,\ldots,f_d)$–adic completions. In fact S/R still will be a locally complete intersection as follows easily from [**KD**], (C.18) and (7.1), and by ([**L$_1$**], (2.3)), (7.1) and trace axiom TR3 (base change) the residues in question don't change, if we identify k with its image in its $\mathfrak{m}$–adic completion. But then R is a direct product of local rings, and the claim follows from the discussion at the beginning of this section.

Suppose now that assumptions ii. or iii. are satisfied. By (7.1), [**L$_1$**](2.3), (4.11) and (6.4) we may replace R and S by their $(f_1,\ldots,f_d)$–adic completions $\hat{R}$ and $\hat{S}$, and we may replace $\Omega_{R/k}$ and $\Omega_{S/k}$ by $\widetilde{\Omega}_{\hat{R}/k}$ and $\widetilde{\Omega}_{\hat{S}/k}$, which exist by (2.6)ii. Note that if $\Omega^1_{R/k}$ is finite and projective as an R–module, then $\widetilde{\Omega}^1_{\hat{R}/k} = \Omega^1_{R/k} \otimes_R \hat{R}$ is finite and projective as an $\hat{R}$–module. For simplicity we will write R resp. S instead of $\hat{R}$ resp. $\hat{S}$.

Now R is a finite flat $k[[\mathbf{X}]] = k[[X_1, \ldots, X_d]]$-algebra via $X_i \mapsto f_i$ by $[\mathbf{L_1}](3.3.2)$, and we have well defined traces of differential forms

$$\widetilde{\sigma}_{S/R} : \widetilde{\Omega}^{\cdot}_{S/k} \to \widetilde{\Omega}^{\cdot}_{R/k}$$

$$\widetilde{\sigma}_{S/k[[\mathbf{X}]]} : \widetilde{\Omega}^{\cdot}_{S/k} \to \widetilde{\Omega}^{\cdot}_{k[[\mathbf{X}]]/k}$$

and

$$\widetilde{\sigma}_{R/k[[\mathbf{X}]]} : \widetilde{\Omega}^{\cdot}_{R/k} \to \widetilde{\Omega}^{\cdot}_{k[[\mathbf{X}]]/k}.$$

(Note that if 2 is not a zerodivisor of R then it is not a zerodivisor of $k[[\mathbf{X}]]$.)

By (6.9) we have for $\widetilde{\omega} \in \widetilde{\Omega}^d_{S/k}$

$$\widetilde{\mathrm{Res}}^d_{S/k} \begin{bmatrix} \widetilde{\omega} \\ f_1, \ldots, f_d \end{bmatrix} = \widetilde{\mathrm{Res}}^d_{k[[\mathbf{X}]]/k} \begin{bmatrix} \widetilde{\sigma}_{S/k[[\mathbf{X}]]}(\widetilde{\omega}) \\ X_1, \ldots, X_d \end{bmatrix}$$

and

$$\widetilde{\mathrm{Res}}^d_{R/k} \begin{bmatrix} \widetilde{\sigma}_{S/R}(\widetilde{\omega}) \\ f_1, \ldots, f_d \end{bmatrix} = \widetilde{\mathrm{Res}}^d_{k[[\mathbf{X}]]/k} \begin{bmatrix} \widetilde{\sigma}_{R/k[[\mathbf{X}]]}(\widetilde{\sigma}_{S/R}(\widetilde{\omega})) \\ X_1, \ldots, X_d \end{bmatrix}.$$

Since in either case $k[[X_1, \ldots, X_d]]$ is reduced and $\widetilde{\Omega}^{\cdot}_{k[[\mathbf{X}]]/k}$ is torsion free we have by (4.13):

$$\widetilde{\sigma}_{S/k[[\mathbf{X}]]} = \widetilde{\sigma}_{R/k[[\mathbf{X}]]} \circ \widetilde{\sigma}_{S/R},$$

and the claim follows.

From the claim in the discrete case trace formula II follows immediately for universally finite differential forms if assumptions i. or ii. are satisfied. If $\widetilde{\Omega}^1_{R/k}$ is a projective R-module, then a slight modification of the proof of $[\mathbf{KD}]$, (12.10) in connection with (2.16) i) shows that $\widetilde{\Omega}^{\cdot}_{\hat{R}/k} = \widetilde{\Omega}^{\cdot}_{R/k} \otimes_R \hat{R}$, where again $\hat{R}$ denotes the $(f_1, \ldots, f_d)$-adic completion of R, and we can complete the proof in this case as above.

7.3. *Remark*: The trace of differential forms constructed in (4.2) coincides with the trace defined by Angéniol in $[\mathbf{A}]$, (7.1.2). Therefore the theorem shows in particular that the trace formula holds true for Angéniol's trace, if the base ring k is reduced.

7.4. *Remark*: The assumption "k is reduced" is only used to ensure the validity of TR5 (transitivity) in the last step of the proof. Therefore if the transitivity of traces can be proved in general, the above theorem will immediately generalize to the case of an arbitrary noetherian k.

7.5. *Remark* ([**L₁**], *app. A*): Suppose that k is a perfect field, and that $(R, \mathfrak{m})$ is a d-dimensional local domain which is a localization of an affine k–algebra, and whose residue field $R/\mathfrak{m}$ is finite over k. Denote by $L = Q(R)$ the field of fractions of R. Then the module $\omega^d_{R/k}$ of regular differential forms is a canonically defined submodule of $\Omega^d_{L/k}$, and the canonical map from ordinary to meromorphic differential forms induces a well–defined morphism

$$c^d_{R/k} : \Omega^d_{R/k} \to \omega^d_{R/k},$$

the so called fundamental class (c.f. [**KW**] §§4,5).

Under the above assumptions a residue morphism

$$\mathrm{res}_R : H^d_{\mathfrak{m}}(\omega^d_{R/k}) \to k$$

can be deduced from duality theory (c.f. [**L₂**]). Let R be a Cohen–Macaulay ring and let $f_1, \ldots, f_d$ be a regular sequence of parameters. Then every element ξ of $H^d_{\mathfrak{m}}(\omega^d_{R/k})$ can be represented as a "generalized fraction"

$$\xi = \omega/(f_1^{m_1}, \ldots, f_d^{m_d})$$

for suitable $\omega \in \omega^d_{R/k}$ and positive integers $m_1, \ldots, m_d$. It is an easy consequence of trace formula II and the work of Lipman ([**L₂**], §9) that

$$\mathrm{res}_R(c^d_{R/k}(\omega)/(f_1^{m_1}, \ldots, f_d^{m_d})) = \mathrm{Res}^d_{R/k} \begin{bmatrix} \omega \\ f_1^{m_1}, \ldots, f_d^{m_d} \end{bmatrix}$$

for every $\omega \in \Omega^d_{R/k}$. Starting with this formula the whole theory of residues on k–varieties can be developed in a very explicit way.

In the remainder of this section it shall be shown that "trace formula II" implies residue axiom (R4) "transitivity" ([**RD**], p.199) under appropriate assumptions. For this we fix the following set up:

Let k be a noetherian ring, and let R/k and S/R be noetherian algebras. Furthermore let $g_1, \ldots, g_d \in R$ and $f_1, \ldots, f_\ell \in S$ be elements such that:

a) $\{g_1, \ldots, g_d\}$ is a quasi–regular sequence in R, and $P := R/(g_1, \ldots, g_d)$ is finite and flat as a k–module.

b) $\{f_1, \ldots, f_\ell\}$ and $\{g_1, \ldots, g_d, f_1, \ldots, f_\ell\}$ are quasi–regular sequences in S and $Q := S/(f_1, \ldots, f_\ell)$ is finite and flat as an R–module.

7.6. LEMMA. *If $Q = S/(f_1, \ldots, f_\ell)$ is a finite locally complete intersection over R, and f S is $(f) = (f_1, \ldots, f_\ell)$–adically complete, then S is a finite locally complete intersection over $R[[Y_1, \ldots, Y_\ell]]$ via $Y_i \mapsto f_i$.*

Proof: By [L₁], (3.3.2) we have that $S/R[[Y_1, \ldots, Y_\ell]]$ is finite and projective. Since $S/(f_1, \ldots, f_\ell)$ is a locally complete intersection over $R = R[[Y_1, \ldots, Y_\ell]]/(Y_1, \ldots, Y_\ell)$, and since $(f_1, \ldots, f_\ell) \subseteq \mathrm{rad}(S)$ and $(Y_1, \ldots, Y_\ell) \subseteq \mathrm{rad}(R[[Y_1, \ldots, Y_\ell]])$ it follows easily from the local criterion of flatness that $S/R[[Y_1, \ldots, Y_\ell]]$ is a locally complete intersection (use [KD], C.7) and (C.4)).

7.7. PROPOSITION. *Suppose that one of the following conditions holds:*

  i. *Q/R is a finite locally complete intersection.*

  ii. *k is a reduced $\mathbf{Q}$–algebra.*

  iii. *k is reduced, 2 is not a zerodivisor of R, and $\Omega^1_{R/k}$ is finite and projective as an R–module.*

Then it holds for $\omega_1 \in \Omega^{d+1}_{R/k}$ and $\omega_2 \in \Omega^{\ell-1}_{S/k}$

$$\mathrm{Res}^{d+\ell}_{S/k} \left[\begin{array}{c} \omega_1' \cdot \omega_2 \\ g_1, \ldots, g_d, \ f_1, \ldots, f_\ell \end{array} \right] = 0$$

where ω_1' denotes the image of ω_1 by the canonical map $\Omega^{\cdot}_{R/k} \to \Omega^{\cdot}_{S/k}$. The same formula holds true for universally finite differential forms if they exist, and if $\Omega^{\cdot}_{R/k} \to \Omega^{\cdot}_{S/k}$ induces $\widetilde{\Omega}^{\cdot}_{R/k} \to \widetilde{\Omega}^{\cdot}_{S/k}$.

Proof: We may replace R by its $(\mathbf{g})$–adic completion $\hat{R}$ and S by its $(\mathbf{g}, \mathbf{f})$–adic completion and $\Omega^{\cdot}_{R/k}$ resp. $\Omega^{\cdot}_{S/k}$ by $\widetilde{\Omega}^{\cdot}_{\hat{R}/k}$ resp $\widetilde{\Omega}^{\cdot}_{\hat{S}/k}$, which exist by (2.6):

In fact if Q/R is a finite locally complete intersection, then so is $\hat{S}/(f_1, \ldots, f_\ell)$ over $\hat{R}$ by (7.1) and [KD], (C.18), and if $\Omega^1_{R/k}$ is finite and projective as an R–module, then so is $\widetilde{\Omega}^1_{\hat{R}/k}$ as an $\hat{R}$–module by [KD], (12.5)b). Furthermore the canonical map $\Omega^{\cdot}_{R/k} \to \Omega^{\cdot}_{S/k}$ induces a homomorphism $\widetilde{\Omega}^{\cdot}_{\hat{R}/k} \to \widetilde{\Omega}^{\cdot}_{\hat{S}/k}$ by the continuity of $R \to S$. Furthermore $\{g_1, \ldots, g_d\} \subseteq \hat{R}$, $\{f_1, \ldots, f_\ell\} \subseteq \hat{S}$ and $\{g_1, \ldots, g_d, \ f_1, \ldots, f_\ell\} \subseteq \hat{S}$ are still quasi–regular sequences (using c.f. [GS], Prop. 6.2.b), and hence they are actually regular by [L₁], (3.2)(a). Therefore it obviously suffices to show:

For $r_1, r_2, \ldots, r_{d+1} \in \hat{R}$ and $s_0, s_1, \ldots, s_{\ell-1} \in \hat{S}$ it holds

$$\widetilde{\mathrm{Res}}_{\hat{S}/k}^{d+\ell} \begin{bmatrix} s_0 dr_1, \ldots, dr_{d+1} ds_1 \ldots ds_{\ell-1} \\ g_1, \ldots, g_d, \ f_1, \ldots, f_\ell \end{bmatrix} = 0.$$

So we have reduced to the following case:

    $-R$ is finite and flat over $k[[X_1, \ldots, X_d]]$ via $X_i \mapsto g_i$.

    $-S$ is finite and flat over $k[[X_1, \ldots, X_d, Y_1, \ldots, Y_\ell]]$, via $X_i \mapsto g_i$, $Y_j \mapsto f_j$, and S is finite and flat over $R[[Y_1, \ldots, Y_\ell]]$ via $Y_j \mapsto f_j$.

    $-\omega_1' \omega_2 = s_0 dr_1 \ldots dr_{d+1} ds_1 \ldots ds_{\ell-1} \in \tilde{\Omega}_{S/k}^{d+\ell}$.

Furthermore we have under assumption i. that $S/R[[Y_1, \ldots Y_\ell]]$ is a finite locally complete intersection (by (7.6)), and under assumption iii. we have that $\tilde{\Omega}_{R[[\mathbf{Y}]]/k}^1$ is finite and projective as an $R[[Y_1, \ldots, Y_\ell]]$–module, and that 2 is not a zerodivisor of $R[[Y_1, \ldots, Y_\ell]] =: R[[\mathbf{Y}]]$.

So in any case we get a well–defined trace:

$$\tilde{\sigma}_{S/R[[\mathbf{Y}]]} : \tilde{\Omega}_{S/k}^{\cdot} \to \tilde{\Omega}_{R[[\mathbf{Y}]]/k}^{\cdot}$$

and it holds by (7.2) and TR1 "Ω–linearity":

$$\widetilde{\mathrm{Res}}_{S/k}^{\ell+d} \begin{bmatrix} s_0 dr_1 \ldots dr_{d+1} ds_1 \ldots dr_{\ell-1} \\ g_1, \ldots, g_d, \ f_1, \ldots, f_\ell \end{bmatrix}$$

$$= \widetilde{\mathrm{Res}}_{R[[\mathbf{Y}]]/k}^{\ell+d} \begin{bmatrix} \tilde{\sigma}_{S/R[[\mathbf{Y}]]}(s_0 dr_1 \ldots dr_{d+1} ds_1 \ldots ds_{\ell-1}) \\ g_1, \ldots, g_d, Y_1, \ldots, Y_\ell \end{bmatrix}$$

$$= \widetilde{\mathrm{Res}}_{R[[\mathbf{Y}]]/k}^{\ell+d} \begin{bmatrix} dr_1 \ldots dr_{d+1} \tilde{\sigma}_{S/R[[\mathbf{Y}]]}(s_0 ds_1 \ldots ds_{\ell-1}) \\ g_1, \ldots, g_d, \ Y_1, \ldots, Y_\ell. \end{bmatrix}$$

Therefore we may even assume $S = R[[Y_1, \ldots Y_\ell]]$ and $f_j = Y_j (j = 1, \ldots, \ell)$.

It is now possible to complete the proof of the proposition by using only the definition in $[L_1]$, §1 and the basic formulas stated there. The calculations involved are straightforward but somewhat lengthy. Therefore we will follow a different approach, using some of the more involved results of Lipman.

There exist canonical homomorphisms

$$R \to \mathrm{Hom}_{k[[\mathbf{X}]]}(R, R), r \mapsto \mu_r^R = \text{"multiplication by } r\text{"}$$

$$S \to \mathrm{Hom}_{k[[\mathbf{X}, \mathbf{Y}]]}(S, S), s \mapsto \mu_s^S = \text{"multiplication by } s\text{"}$$

nd under the above assumptions we get a commutative diagram

$$
\begin{array}{ccc}
S = R[[\mathbf{Y}]] & \xrightarrow{\ \mu^S\ } & \mathrm{Hom}_{k[[\mathbf{X},\mathbf{Y}]]}(S,S) = E[[\mathbf{Y}]] \\
{\scriptstyle\text{can}}\ \Big\uparrow & & \Big\uparrow\ \Psi \\
R & \xrightarrow{\ \mu^R\ } & \mathrm{Hom}_{k[[\mathbf{X}]]}(R,R) =: E,
\end{array}
$$

here Ψ is defined as follows:

For $\varphi \in E$ and $s = \Sigma r_\mu \mathbf{Y}^\mu \in R[[\mathbf{Y}]] = S$ set

$$
\Psi(\varphi)(s) := \Sigma \varphi(r_\mu) \mathbf{Y}^\mu .
$$

fact since S is a power series ring over R we have for $r \in R$ and $s = \Sigma r_\mu \mathbf{Y}^\mu \in S$

$$
\mu_r^S(s) = \Sigma \mu_r^R(r_\mu) \mathbf{Y}^\mu
$$

plying the commutativity of the above diagram. Hence by the identification $\mathrm{Hom}_{k[[\mathbf{X},\mathbf{Y}]]}(S,S) =$ [[Y]], induced by the canonical map $R \to S$ which is a section of the canonical homomor- ism $S \to S/(Y_1,\ldots,Y_\ell) = R$, we have:

$$
\mu_{r_i}^S \in E \subseteq E[[\mathbf{Y}]] \quad \text{for } i = 1,\ldots,d+1.
$$

herefore it holds
$$
\frac{\partial}{\partial Y_j}(\mu_{r_i}^S) = 0 \quad \text{for } i = 1,\ldots,d+1;\ j = 1,\ldots,\ell.
$$

he claim of the proposition for absolute differential forms follows from [$\mathbf{L_1}$], (3.7) (in connec- on with 6.4)) by an easy exercise in determinants. For universally finite differential forms e proof can be completed by arguing as in the proof of (7.2).

7.8. THEOREM. *Suppose that one of the following conditions holds:*

i. $Q = S/(f_1,\ldots,f_\ell)$ *is a finitely locally complete intersection over* R.

ii. k *is a reduced* Q–*algebra.*

iii. k *is reduced, 2 is not a zerodivisor of* R, *and* $\Omega^1_{R/k}$ *is finite and projective as an* -module.

Then it holds for $\omega_1 \in \Omega^d_{R/k}$ and $\omega_2 \in \Omega^\ell_{S/k}$:

$$\operatorname{Res}^{d+\ell}_{S/k} \begin{bmatrix} \omega_1' \cdot \omega_2 \\ g_1, \ldots, g_d, \ f_1, \ldots, f_\ell \end{bmatrix} = \operatorname{Res}^d_{R/k} \begin{bmatrix} \omega_1 \operatorname{Res}^\ell_{S/R} \begin{bmatrix} \omega_2'' \\ f_1, \ldots, f_\ell \end{bmatrix} \\ g_1, \ldots, g_d \end{bmatrix}$$

where ω_1' denotes the image of ω_1 by $\Omega^{\cdot}_{R/k} \to \Omega^{\cdot}_{S/k}$, and where ω_2'' denotes the image of ω_2 by $\Omega^{\cdot}_{S/k} \to \Omega^{\cdot}_{S/R}$.

If the universally finite differential algebras $\widetilde{\Omega}^{\cdot}_{R/k}$ of R/k and $\widetilde{\Omega}^{\cdot}_{S/k}$ of S/k exist, and if $\Omega^{\cdot}_{R/k} \to \Omega^{\cdot}_{S/k}$ induces a morphism $\widetilde{\Omega}^{\cdot}_{R/k} \to \widetilde{\Omega}^{\cdot}_{S/k}$, then the same formula holds true for universally finite differential forms.

Before proving (7.8), we state some consequences.

7.9. COROLLARY ((R4) "TRANSITIVITY", [RD], P. 199). *Suppose that R/k is essentially of finite type, smooth, and equidimensional of dimension d, and that S/R is essentially of finite type, smooth and equidimensional of dimension ℓ. Assume that $g_1, \ldots, g_d \in R$ and $f_1, \ldots, f_\ell \in S$ are elements such that $R/(g_1, \ldots, g_d)$ is a finite k–module and $S/(f_1, \ldots, f_\ell)$ is a finite R–module.*

Then it holds for $\omega_1 \otimes \omega_2 \in \Omega^d_{R/k} \otimes_R \Omega^\ell_{S/R} \cong \Omega^{d+\ell}_{S/k}$:

$$\operatorname{Res}^{d+\ell}_{S/k} \begin{bmatrix} \omega_1 \otimes \omega_2 \\ g_1, \ldots, g_d, \ f_1, \ldots, f_\ell \end{bmatrix} = \operatorname{Res}^d_{R/k} \begin{bmatrix} \omega_1 \operatorname{Res}^\ell_{S/R} \begin{bmatrix} \omega_2 \\ f_1, \ldots, f_\ell \end{bmatrix} \\ g_1, \ldots, g_d \end{bmatrix}$$

Proof: Under the present assumptions $\{g_1, \ldots, g_d\} \subseteq R$, $\{f_1, \ldots, f_\ell\} \subseteq S$ and $\{g_1, \ldots, g_d, \ f_1, \ldots, f_\ell\} \subseteq S$ are quasi–regular sequences, and $R/(g_1, \ldots, g_d)$ is finite and flat over k, and $S/(f_1, \ldots, f_\ell)$ is finite, flat and a relative complete intersection over R.

7.10. COROLLARY. *Suppose R/k and S/R are power series algebras of relative dimension d and ℓ, and suppose $g_1, \ldots, g_d \in R$ and $f_1, \ldots, f_\ell \in S$ are elements such that $R/(g_1, \ldots, g_d)$ is a finite k–module, and such that $S/(f_1, \ldots, f_\ell)$ is a finite R–module. Then it holds for $\omega_1 \otimes \omega_2 \in \widetilde{\Omega}^d_{R/k} \otimes_R \widetilde{\Omega}^\ell_{S/R} \cong \widetilde{\Omega}^{d+\ell}_{S/k}$:*

$$\widetilde{\operatorname{Res}}^{d+\ell}_{S/k} \begin{bmatrix} \omega_1 \otimes \omega_2 \\ g_1, \ldots, g_d, \ f_1, \ldots, f_\ell \end{bmatrix} = \widetilde{\operatorname{Res}}^d_{R/k} \begin{bmatrix} \omega_1 \widetilde{\operatorname{Res}}^\ell_{S/R} \begin{bmatrix} \omega_2 \\ f_1, \ldots, f_\ell \end{bmatrix} \\ g_1, \ldots, g_d \end{bmatrix}.$$

Proof: As the proof of (7.9).

Proof of theorem (7.8): As in the proof of (7.7) we may replace R by its $(\mathbf{g})$–adic completion and S by its $(\mathbf{g}, \mathbf{f})$–adic completion, and it also suffices to prove the formula for universally finite differential forms. As above we therefore have that S is finite and projective as an $R[[\mathbf{Y}]] = R[[Y_1, \ldots, Y_\ell]]$–module via $Y_j \mapsto f_j$.

There exist well defined trace maps

$$\tilde{\sigma}^1_{S/R[[\mathbf{Y}]]} : \tilde{\Omega}^{\cdot}_{S/k} \to \tilde{\Omega}^{\cdot}_{R[[\mathbf{Y}]]/k}$$

and

$$\tilde{\sigma}^2_{S/R[[\mathbf{Y}]]} : \tilde{\Omega}^{\cdot}_{S/R} \to \tilde{\Omega}^{\cdot}_{R[[\mathbf{Y}]]/R} .$$

Furthermore $\tilde{\Omega}^\ell_{R[[\mathbf{Y}]]/R}$ is free with basis $dY_1 \ldots dY_\ell$, and therefore it holds:

$$\tilde{\sigma}^2_{S/R[[\mathbf{Y}]]}(\omega''_2) = t dY_1 \ldots dY_\ell \text{ for some } t = \Sigma t_\mu \mathbf{Y}^\mu \in R[[\mathbf{Y}]].$$

By trace formula II (7.2) we have:

$$\widetilde{\text{Res}}^d_{R/k} \begin{bmatrix} \omega_1 \cdot \widetilde{\text{Res}}_{S/R} \begin{bmatrix} \omega''_2 \\ f_1, \ldots, f_\ell \\ g_1, \ldots, g_d \end{bmatrix} \end{bmatrix}$$

$$= \widetilde{\text{Res}}^d_{R/k} \begin{bmatrix} \omega_1 \cdot \widetilde{\text{Res}}^\ell_{R[[\mathbf{Y}]]/R} \begin{bmatrix} \tilde{\sigma}^2_{S/R[[\mathbf{Y}]]}(\omega''_2) \\ Y_1, \ldots, Y_\ell \\ g_1, \ldots, g_d \end{bmatrix} \end{bmatrix}$$

$$= \widetilde{\text{Res}}^d_{R/k} \begin{bmatrix} \omega_1 \widetilde{\text{Res}}^\ell_{R[[\mathbf{Y}]]/R} \begin{bmatrix} t dY_1 \ldots dY_\ell \\ Y_1, \ldots, Y_\ell \\ g_1, \ldots, g_d \end{bmatrix} \end{bmatrix}$$

$$= \widetilde{\text{Res}}^d_{R/k} \begin{bmatrix} t_0 \omega_1 \\ g_1, \ldots, g_d \end{bmatrix}$$

where t_0 is the constant term of $t \in R[[\mathbf{Y}]]$ (see (6.12)). Similarly it holds:

$$\widetilde{\text{Res}}^{d+\ell}_{S/k} \begin{bmatrix} \omega'_1 \cdot \omega_2 \\ g_1, \ldots, g_d, f_1, \ldots, f_\ell \end{bmatrix}$$

$$= \widetilde{\text{Res}}^{d+\ell}_{R[[\mathbf{Y}]]/k} \begin{bmatrix} \tilde{\sigma}^1_{S/R[[\mathbf{Y}]]}(\omega'_1 \cdot \omega_2) \\ g_1, \ldots, g_d, Y_1, \ldots, Y_\ell \end{bmatrix}$$

$$= \widetilde{\text{Res}}^{d+\ell}_{R[[\mathbf{Y}]]/k} \begin{bmatrix} \omega'_1 \tilde{\sigma}^1_{S/R[[\mathbf{Y}]]}(\omega_2) \\ g_1, \ldots, g_d, Y_1, \ldots, Y_\ell \end{bmatrix} .$$

Claim:

$$\widetilde{\operatorname{Res}}_{R[[Y]]/k}^{d+\ell}\begin{bmatrix} \omega_1'\widetilde{\sigma}_{S/R[[Y]]}(\omega_2) \\ g_1,\ldots,g_d,\ Y_1,\ldots,Y_\ell \end{bmatrix}$$
$$=\widetilde{\operatorname{Res}}_{R[[Y]]/k}^{d+\ell}\begin{bmatrix} \omega_1'\cdot tdY_1\ldots dY_\ell \\ g_1,\ldots,g_d,\ Y_1,\ldots,Y_\ell \end{bmatrix}.$$

Proof of the claim: By (TR3) "base change" we have that the image of $\widetilde{\sigma}_{S/R[[Y]]}^1(\omega_2)$ by the canonical map $\widetilde{\Omega}_{R[[Y]]/k}^\ell \to \widetilde{\Omega}_{R[[Y]]/R}^\ell$ is equal to $t\cdot dY_1\ldots dY_\ell$, and therefore it follows:

$$t\cdot dY_1\ldots dY_\ell - \widetilde{\sigma}_{S/R[[Y]]}^1(\omega_2) \in \ker(\widetilde{\Omega}_{R[[Y]]/k}^\ell \to \widetilde{\Omega}_{R[[Y]]/R}^\ell) \qquad \text{(by } [\mathbf{KD}],\ (11.7),)$$
$$=(\{dr : r \in R\})$$

i.e. we have

$$\omega_1'\cdot(t\cdot dY_1\ldots dY_\ell - \widetilde{\sigma}_{S/R[[Y]]}^1(\omega_2)) = \Sigma\eta_i'\tau_i$$
$$\text{with } \eta_i \in \widetilde{\Omega}_{R/k}^{d+1} \text{ and } \tau_i \in \Omega_{R[[Y]]/k}^{\ell-1}.$$

Hence by (7.7):

$$\widetilde{\operatorname{Res}}_{R[[Y]]/k}^{d+\ell}\begin{bmatrix} \omega_1'(tdY_1\ldots dY_\ell - \widetilde{\sigma}_{S/R[[Y]]}^1(\omega_2)) \\ g_1,\ldots,g_d,\ Y_1,\ldots,Y_\ell \end{bmatrix} = 0,$$

and the claim follows.

To complete the proof of (7.8) for absolute differential forms it remains to note that

$$\widetilde{\operatorname{Res}}_{R/k}^d\begin{bmatrix} t_0\omega_1 \\ g_1,\ldots,g_d \end{bmatrix} = \widetilde{\operatorname{Res}}_{R[[Y]]/k}^{d+\ell}\begin{bmatrix} \omega_1'tdY_1\ldots dY_\ell \\ g_1,\ldots,g_d,\ Y_1,\ldots,Y_\ell \end{bmatrix}$$

which is an immediate consequence of $[\mathbf{L_1}]$, (1.10.7). For universally finite differential forms the proof can be completed by arguing as in the proof of (7.2).

7.11. Remark: In the proofs of (7.7) and (7.8) the assumption "k is reduced" is only needed to be able to apply "trace formula II" (7.2). So again if the transitivity of traces can be proved in general, the above results will immediately generalize to the case of an arbitrary noetherian k.

7.12. Remark: If R is reduced and noetherian, if Ω is an exterior differential algebra of R such that Ω^1 is a finite projective R-module, and if S is a finite flat R-algebra, then by (4.4) there also exists a well defined trace

$$\sigma_{S/R} : \Omega_S \to \Omega.$$

Hence in (7.2), (7.7) and (7.8) in assumption iii. the condition "2 is not a zerodivisor of R" can be replaced by appropriate conditions ensuring that the $(\mathbf{g})$–adic completion $\hat{R}$ of R is reduced, for example by the condition "R is reduced and excellent" ([**Mat**], Thm. 79).

In particular if k is a field and R is the local ring of a smooth closed point of a k–variety X, then the assertions of (7.2), (7.7) and (7.8) hold, even if char$(k) = 2$.

REFERENCES

[A] ANGÉNIOL, B., *Familles de cycles algébriques–Schéma de Chow*, Springer Lecture N in Mathematics **896** (1981).

[AL] ANGÉNIOL, B.; LEJEUNE–JALABERT, M., *Calcul différentiel et classes charactérist en géometrie algébrique*, Prepublication de L'Institut Fourier **28** (1985).

[AM] ATIYAH, M. F.; MACDONALD, J. G., "Introduction to Commutative Algebra," Add Wesley, Reading, 1969.

[B₁] BOURBAKI, N., "Commutative Algebra," Hermann, Paris, 1972.

[B₂] BOURBAKI, N., "General Topology, part I," Hermann, Paris, 1966.

[B₃] BOURBAKI, N., "Algebra, part I," Hermann, Paris, 1974.

[EH] EAGON, J. A.; HOCHSTER, M., *Cohen–Macaulay rings, invariant theory and the ge* *perfection of determinantal loci*, American Journal of Mathematics **93** (1971), 1020–1

[EGA I] GROTHENDIECK, A.; DIEUDONNÉ, J., "Eléments de Géometrie algébrique I," Sprir Berlin, Heidelberg, New York, 1971.

EGA II–IV] GROTHENDIECK, A.; DIEUDONNÉ, J., *Eléments de géometrie algébrique*, Publ. N IHES **8** (1961), **11** (1961), **17** (1963), **20** (1964), **24** (1965), **28** (1966), **32** (1967).

[GS] GRECO, S.; SALMON, P., "Topics in m-adic Topologies," Springer, Berlin, Heidelb New York, 1971.

[Ha] HARTSHORNE, R., "Algebraic Geometry," Springer, Berlin, Heidelberg, New York, 1

[Ho] HOCHSCHILD, G., *Relative homological algebra*, Transactions Amer. Math. Soc (1956), 246–269.

[HKR] HOCHSCHILD, G.; KOSTANT, B.; ROSENBERG, A., *Differential forms of regular a* *algebras*, Transactions Amer. Math. Soc. **102** (1962), 383–408.

[Hü] HÜBL, R., *Spuren von Differentialformen und Hochschild–Homologie*, Dissertation (1 Regensburg.

[KD] KUNZ, E., "Kähler Differentials," Vieweg, Braunschweig, Wiesbaden, 1986.

[K₁] KUNZ, E., "Einführung in die kommutative Algebra und algebraische Geometrie," Vieweg, Braunschweig Wiesbaden, 1979.

[K₂] KUNZ, E., *Lectures on Kähler Differentials, part II*, Manuscript (1985), Regensburg.

[K₃] KUNZ, E., *Differentialformen inseparabler algebraischer Funktionenkörper*, Math. Zeitschr. **76** (1961), 56–74.

[K₄] KUNZ, E., *Arithmetische Anwendungen der Differentialalgebren*, Journal f. d. reine u. angew. Mathematik **214/215** (1964), 276–320.

[KW] KUNZ, E.; WALDI, R., *Regular differential forms*. To appear in: Contemporary Mathematics.

[L₁] LIPMAN, J., *Residues and traces of differential forms via Hochschild homology*, Contemporary Mathematics **61** (1987), Amer. Math. Soc., Providence.

[L₂] LIPMAN, J., *Dualizing sheaves, differentials and residues on algebraic varieties*, Astérisque **117** (1984).

[Ma] MATUSUMURA, H., "Commutative Algebra (sec. ed.)," Benjamin, Reading, 1980.

[ML] MACLANE, S., "Homology," Springer, Berlin, Heidelberg, New York, 1967.

[N] NAGATA, M., "Local Rings," Interscience Publishers, New York, 1962.

[Na] NASTOLD, H. J., *Zum Dualitätssatz in inseparablen Funktionenkörpern der Dimension 1*, Math. Zeitschr. **76** (1961), 75–84.

[R] RINEHART, G. S., *Differential forms of general commutative algebras*, Transactions Amer. Math. Soc. **108** (1963), 195–222.

[RD] HARTSHORNE, R., *Residues and Duality*, Springer Lecture Notes in Mathematics **20** (1966).

[Sal] SALMON, P., *Sur les séries formelles restreintes*, Bul. Soc. Math. France **92** (1964), 385–410.

[Sch] SCHUBERT, H., "Categories," Springer, Berlin, Heidelberg, New York, 1972.

[S] SERRE, J. P., *Algébre Locale–Multiplicités*, Springer Lecture Notes in Mathematics **11** (1965).

Symbol Index

The numbers indicate on which page the symbol is defined or where a reference for its definition is given.

Subject Index

LECTURE NOTES IN MATHEMATICS
Edited by A. Dold and B. Eckmann

Some general remarks on the publication of monographs and seminars

In what follows all references to monographs, are applicable also to multiauthorship volumes such as seminar notes.

§1. Lecture Notes aim to report new developments - quickly, informally, and at a high level. Monograph manuscripts should be reasonably self-contained and rounded off. Thus they may, and often will, present not only results of the author but also related work by other people. Furthermore, the manuscripts should provide sufficient motivation, examples and applications. This clearly distinguishes Lecture Notes manuscripts from journal articles which normally are very concise. Articles intended for a journal but too long to be accepted by most journals, usually do not have this "lecture notes" character. For similar reasons it is unusual for Ph.D. theses to be accepted for the Lecture Notes series.

Experience has shown that English language manuscripts achieve a much wider distribution.

§2. Manuscripts or plans for Lecture Notes volumes should be submitted either to one of the series editors or to Springer-Verlag, Heidelberg. These proposals are then refereed. A final decision concerning publication can only be made on the basis of the complete manuscripts, but a preliminary decision can usually be based on partial information: a fairly detailed outline describing the planned contents of each chapter, and an indication of the estimated length, a bibliography, and one or two sample chapters - or a first draft of the manuscript. The editors will try to make the preliminary decision as definite as they can on the basis of the available information.

§3. Lecture Notes are printed by photo-offset from typed copy delivered in camera-ready form by the authors. Springer-Verlag provides technical instructions for the preparation of manuscripts, and will also, on request, supply special stationery on which the prescribed typing area is outlined. Careful preparation of the manuscripts will help keep production time short and ensure satisfactory appearance of the finished book. Running titles are not required; if however they are considered necessary, they should be uniform in appearance. We generally advise authors not to start having their final manuscripts specially tpyed beforehand. For professionally typed manuscripts, prepared on the special stationery according to our instructions, Springer-Verlag will, if necessary, contribute towards the typing costs at a fixed rate.

The actual production of a Lecture Notes volume takes 6-8 weeks.

.../...

§4. Final manuscripts should contain at least 100 pages of mathematical text and should include
- a table of contents
- an informative introduction, perhaps with some historical remarks. It should be accessible to a reader not particularly familiar with the topic treated.
- a subject index; this is almost always genuinely helpful for the reader.

§5. Authors receive a total of 50 free copies of their volume, but no royalties. They are entitled to purchase further copies of their book for their personal use at a discount of 33.3 %, other Springer mathematics books at a discount of 20 % directly from Springer-Verlag.

Commitment to publish is made by letter of intent rather than by signing a formal contract. Springer-Verlag secures the copyright for each volume.